Fractal Attraction™

*A Fractal Design System for the **Macintosh**®*

Fractal Attraction™

A Fractal Design System for the ***Macintosh***®

KEVIN D. LEE

College of St. Catherine
St. Paul, Minnesota

and

Sandpiper Software Inc.
St. Paul, Minnesota

YOSEF COHEN

University of Minnesota
St. Paul, Minnesota

ACADEMIC PRESS

Harcourt Brace Jovanovich, Publishers

Boston San Diego New York
London Sydney Tokyo Toronto

Produced in the United States of America.

ISBN 0-12-440740-4

Table of Contents

LIMITED WARRANTY AND DISCLAIMER OF LIABILITY

ACADEMIC PRESS, INC. ("AP"), SANDPIPER SOFTWARE, INC. ("SANDPIPER"), AND ANYONE ELSE WHO HAS BEEN INVOLVED IN THE CREATION OR PRODUCTION OF THE ACCOMPANYING SOFTWARE AND MANUAL (THE "PRODUCT") CANNOT AND DO NOT WARRANT THE PERFORMANCE OR RESULTS THAT MAY BE OBTAINED BY USING THE PRODUCT. THE PRODUCT IS SOLD "AS IS" WITHOUT WARRANTY OF ANY KIND (EXCEPT AS HEREAFTER DESCRIBED), EITHER EXPRESSED OR IMPLIED, INCLUDING, BUT NOT LIMITED TO, ANY WARRANTY OF PERFORMANCE OR ANY IMPLIED WARRANTY OF MERCHANTABILITY OR FITNESS FOR ANY PARTICULAR PURPOSE. AP AND SANDPIPER WARRANT ONLY THAT THE MAGNETIC DISKETTE(S) ON WHICH THE SOFTWARE PROGRAM IS RECORDED IS FREE FROM DEFECTS IN MATERIAL AND FAULTY WORKMANSHIP UNDER NORMAL USE AND SERVICE FOR A PERIOD OF NINETY (90) DAYS FROM THE DATE THE PRODUCT IS DELIVERED. THE PURCHASER'S SOLE AND EXCLUSIVE REMEDY IN THE EVENT OF A DEFECT IS EXPRESSLY LIMITED TO EITHER REPLACEMENT OF THE DISKETTE(S) OR REFUND OF THE PURCHASE PRICE, AT AP'S AND SANDPIPER'S SOLE DISCRETION.

IN NO EVENT, WHETHER AS A RESULT OF BREACH OF CONTRACT, WARRANTY OR TORT (INCLUDING NEGLIGENCE), WILL AP OR SANDPIPER BE LIABLE TO PURCHASER FOR ANY DAMAGES, INCLUDING ANY LOST PROFITS, LOST SAVINGS OR OTHER INCIDENTAL OR CONSEQUENTIAL DAMAGES ARISING OUT OF THE USE OR INABILITY TO USE THE PRODUCT OR ANY MODIFICATIONS THEREOF, OR DUE TO THE CONTENTS OF THE SOFTWARE PROGRAM, EVEN IF AP OR SANDPIPER HAS BEEN ADVISED OF THE POSSIBILITY OF SUCH DAMAGES, OR FOR ANY CLAIM BY ANY OTHER PARTY.

Any request for replacement of a defective diskette must be postage prepaid and must be accompanied by the original defective diskette, your mailing address and telephone number, and proof of date of purchase and purchase price. Send such requests to Sandpiper Software, P.O. Box 8012, St. Paul, MN 55108. AP and Sandpiper shall have no obligation to refund the purchase price or to replace a diskette based on claims of defects in the nature or operation of the Product.

Some states do not allow limitation on how long an implied warranty lasts, not exclusions or limitations of incidental or consequential damages, so the above limitations and exclusions may not apply to you. This Warranty gives you specific legal rights, and you may also have other rights which vary from jurisdiction to jurisdiction.

THE RE-EXPORT OF UNITED STATES ORIGIN SOFTWARE IS SUBJECT TO THE UNITED STATES LAWS UNDER THE EXPORT ADMINISTRATION ACT OF 1969 AS AMENDED. ANY FURTHER SALE OF THE PRODUCT SHALL BE IN COMPLIANCE WITH THE UNITED STATES DEPARTMENT OF COMMERCE ADMINISTRATION REGULATIONS. COMPLIANCE WITH SUCH REGULATIONS IS YOUR RESPONSIBILITY AND NOT THE RESPONSIBILITY OF AP OR SANDPIPER.

§1. What is Fractal Attraction?

Fractal Attraction is a Macintosh application with which you use Iterated Function Systems (IFS) to design your own fractals. As you will see, the ideas behind IFS, which were developed by Barnsley (1988) and co-workers are surprisingly simple, and yet, can produce pictures of incredible complexity and beauty.

The following pages first give a quick overview to using *Fractal Attraction*, then explain what IFS is, and what it does, and show you how to use *Fractal Attraction.* If you are familiar with IFS, you can certainly skip the appropriate sections (e.g. Mathematical Background), you may, however, want to skim through the sections which explain how to design your own fractals. If you are familiar with both IFS and the Macintosh you may want to skip the manual altogether, although *it is suggested that you read the sections on using the **Design** window, the collage example and, if you are using a color Mac, the section on color rendering.*

No apologies are made for the very basic and intuitive explication of the ideas. Throughout it is assumed that you are familiar with basic Macintosh concepts such as clicking, dragging, and menus.

§1.1. Hardware and System Software Required

Two versions of *Fractal Attraction* are included on the master disk, standard and enhanced. The two versions are functionally equivalent except for speed, the enhanced version takes advantage of the math coprocessor and the faster processor on the newer machines. The standard version will run on any Macintosh from the Plus up, but it is intended for use only with the Plus,SE, or LC (or Mac IIsi if it does not have the optional math coprocessor). If you own an SE or Plus rest assured that you can still enjoy *Fractal Attraction* , much time was spent optimizing the standard version for speed.

If you are using the standard version on the LC or IIsi, it will automatically adjust to be in color.

It is suggested that you use system software 6.0 or later.

§1.2. Installing Fractal Attraction

To install *Fractal Attraction* simply drag the appropriate version of the application for your machine and the two the folders named IFS Examples and Initial Figures onto your hard disk. For a floppy disk system copy to a floppy disk.

§1.3. A Quick Tour

To quickly introduce you to the capabilities of *Fractal Attraction* we will create the so-called Sierpinski triangle. Hopefully this will encourage you to read further and find out why the twenty-one numbers that constitute our IFS example produce the Sierpinski triangle.

Start a session with *Fractal Attraction* by double-clicking on the *Fractal Attraction* icon. Three windows will appear. They are titled **Untitled** (no pun intended), **Design**, and **Fractal.** The **Untitled** window is where you enter and edit the IFS code in a spreadsheet like fashion; we will refer to this window as the **code** window. The title will change to match the name of the file. The **design** window also allows you to edit the IFS code but in a draw/paint like fashion. Using this window you can quickly and easily design a fractal. The final window, **Fractal,** is where the image specified by the IFS is plotted.

To illustrate the ease with which *Fractal Attraction* lets you design fractals we will start with the **design** window. Make the **design** window the active window either by clicking on it or choosing **Design** under the **Fractal** menu. You will notice two rectangles, a large one with a dotted outline and a smaller one inside the large one (see figure 1). This smaller one is actually a transformation (reduced copy) of the large rectangle (the result of an equation, but you can read about that later). Fractal Attraction automatically supplies one transformation. In the **design** window you will add two more to create the Siepinski triangle.

First move the first transformation to a new location, click on any edge of the small rectangle, it now becomes selected. Click again inside the figure but this time keep pressing the button and drag the rectangle to the position shown in figure 1.2.

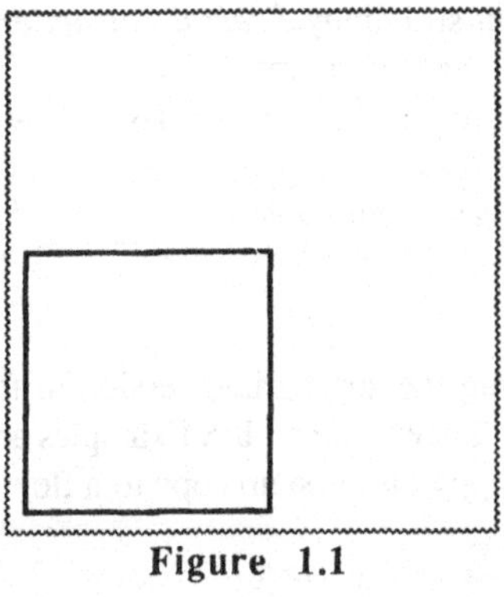

Figure 1.1

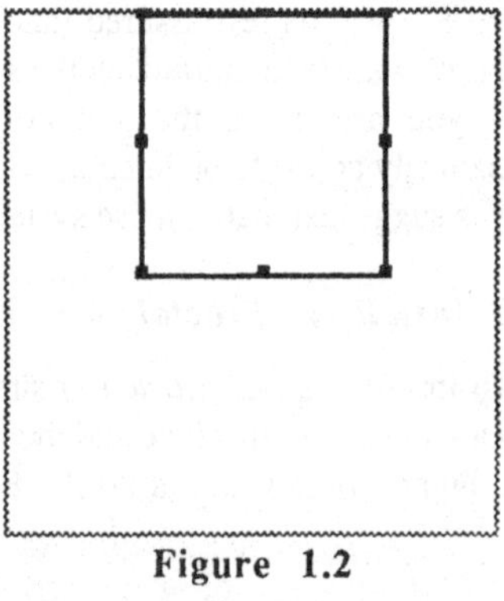

Figure 1.2

Now add a new transformation by selecting **New Transformation** under the **design** menu (see figure 1.3). Using your moving skills, position this new transformation as shown in figure 1.4.

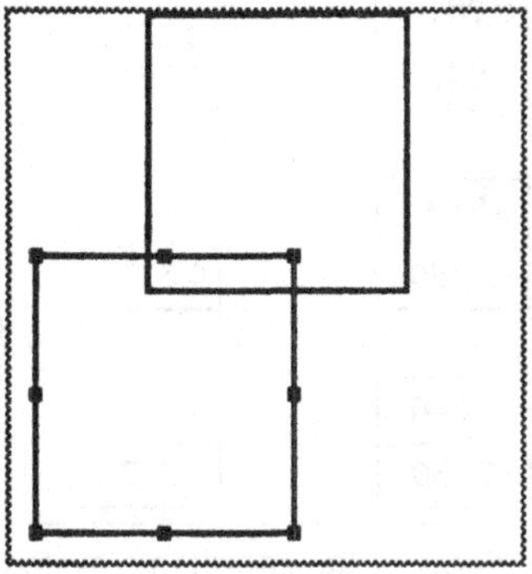

Figure 1.3

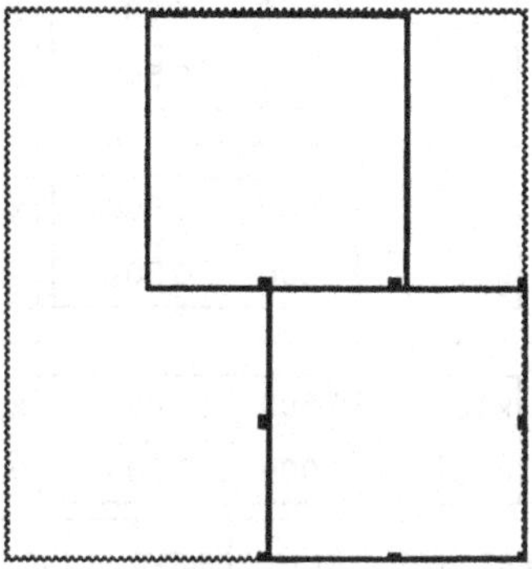

Figure 1.4

Add one more transformation. Position it as shown in figure 1.5. You have now built an IFS code (system of equations, i.e. the transformations) for generating a Sierpinski triangle.

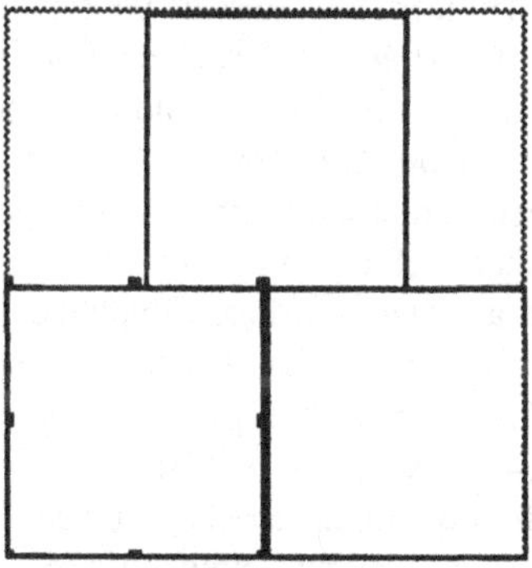

Figure 1.5

Working in the Code Window: Make the **code** window the active window and observe the IFS code you created. Your numbers will probably be slightly different from the numbers shown here. In this window you can edit the parameters (numbers) by pointing and clicking. Most of the time you will do your creating in the **design** window and use the **code** window for fine tuning. By the way, you have now specified a set of 3 equations which act on points in an initial picture (to be explained later).

3 Number of Transformations [1 to 15]

	Transformation	Probability
1.	$\begin{bmatrix} x_{new} \\ y_{new} \end{bmatrix} = \begin{bmatrix} 0.500 & 0.000 \\ 0.000 & 0.500 \end{bmatrix} \begin{bmatrix} x_{old} \\ y_{old} \end{bmatrix} + \begin{bmatrix} 0.281 \\ 0.548 \end{bmatrix}$	0.333
2.	$\begin{bmatrix} x_{new} \\ y_{new} \end{bmatrix} = \begin{bmatrix} 0.500 & 0.000 \\ 0.000 & 0.500 \end{bmatrix} \begin{bmatrix} x_{old} \\ y_{old} \end{bmatrix} + \begin{bmatrix} 0.559 \\ -0.050 \end{bmatrix}$	0.333
3.	$\begin{bmatrix} x_{new} \\ y_{new} \end{bmatrix} = \begin{bmatrix} 0.500 & 0.000 \\ 0.000 & 0.500 \end{bmatrix} \begin{bmatrix} x_{old} \\ y_{old} \end{bmatrix} + \begin{bmatrix} -0.046 \\ -0.050 \end{bmatrix}$	0.333

The next item to observe in the **code** window is the coordinate system; that is the maximum and minimum values for the x and y in the picture to be produced. By default, *Fractal Attraction* scales the coordinate system for you (you will learn how to scale your own coordinates later). Scroll to the bottom of the **Untitled** window. You should see four cells labeled ***x min***, ***x max***, ***y min***, and ***y max***. They should be set to -0.1, 1.1, -0.1 and 1.1, respectively.

Plotting the Image in the Fractal Window: You may now proceed to plot the Sierpinski triangle. Bring the Fractal window to the front by either clicking on it or by choosing Fractal from under the Window menu. There are two algorithms implemented for plotting the fractal (that is, the picture specified by the equations): random and deterministic.

The random algorithm is faster—we will use it. Make sure **Random** is chosen (checked) under the **Fractal** menu. Now choose **Do 10,000**, also under the **Fractal** menu. *Fractal Attraction* is now drawing in the **Fractal** window. Eventually you should see the Sierpinski triangle as shown here. If you want a crisper image, choose **Do 10,000** again.

You may be wondering about the numbers in the bottom of the **fractal** window. Move the mouse over the fractal image and watch how the numbers change. The first pair of numbers tells you the *x* and *y* location of the mouse in the current coordinate system. The third number tells you how many times the point has been 'hit' (explained later). The fourth number tells how many iterations (more about iterations later) have been calculated. The final number is the number of seconds it took to generate the last set of iterations.

More of the Design Window.

This is a good time to show off the power of *Fractal Attraction.* Make the **design** window the active window. Click on the the top rectangle, it will become selected and eight very small rectangles (handles) will appear. While pressing the option key click and drag one of the corner handles. This allows you to rotate the transformation, try rotating 45 degrees (see figure 1.6). Now make the **fractal** window active, and try running another 10000 points. (you may want to choose **clear** under the **edit** menu first). Did you get a fractal similar to figure 1.7? In the **design** window you can shrink, expand, rotate, flip and shear your transformation. Section 4, "Using the Design Window", explains all of these features in detail.

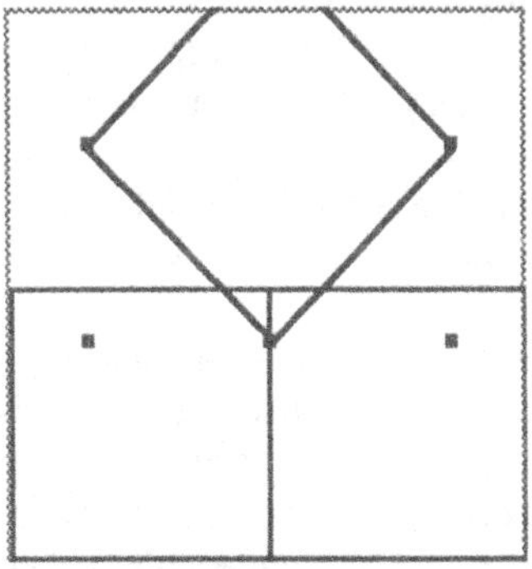

Figure 1.6

Figure 1.7

Zooming In: Once you have generated a fractal you can zoom in on a portion of it.

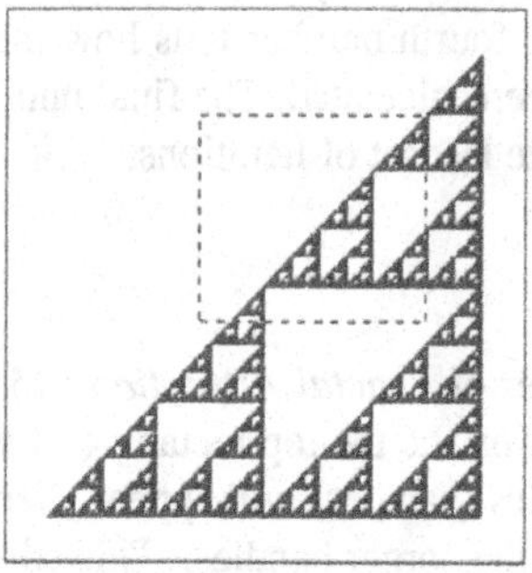

Before **Zoom In**

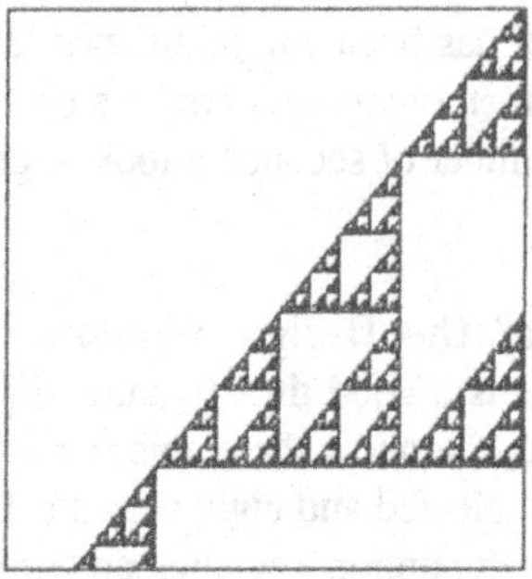

After **Zoom In**

To zoom in, select a portion of the fractal using the mouse (left drawing above). Choose **Zoom In** from under the **Fractal** menu. This will rescale the coordinate system so that the lower left and upper right coordinates of the selection rectangle now become the lower left and upper right coordinates of the window. The image looks rough (right drawing above) until you run the algorithm again at the higher magnification. Try it: run the algorithm for another 10,000 iterations. Do you see the self-similarity property of this picture? This self similarity is one of the distinguishing characteristics of fractals.

Saving and Loading an IFS Code: Make the **Untitled** window the active window. You can save this window by choosing **Save** under the **File** menu.

There are several example files in the folder titled IFS Examples. To access a file (which contains the IFS code) you must first close the active **code** window by choosing **Close** under the **File** menu or clicking in the close box of the window. After closing the **code** window (file), you can use the **Open** command under the **File** menu to look at another file. Choose another file and try 10,000 iterations (after making the **Fractal** window active, and making sure **Random** is checked). You may want to clear the **Fractal** window first by choosing **Clear** under the **Edit** menu.

This completes your quick introduction to *Fractal Attraction*. Hopefully you will find constructing IFS codes and generating images as fascinating as we have. To understand the theory behind the images and more details about the *Fractal Attraction* application read the next section, Mathematical Background. If you already have a background in transformational geometry and IFS codes you may want to skip to the section 4, "Using the **Design** Window".

§2. Mathematical Background

This section will explain the ideas and principles behind generating fractals based on sets of equations. The presentation is at a very basic and intuitive level, and as you will hopefully find out for yourself, does not require sophisticated mathematics.

§2.1. Affine Transformations

We shall start from the very beginning. Our discussion will be limited to the familiar *spaces* we are all used to: points, lines, planes, and so on. These are called *Euclidean* spaces. Consider a single point which constitutes a space. Since the point is all of the space, we need not describe its position. We then say that the *dimension* of a single-point space is 0. Next, consider a line, with an origin and a direction (Figure 2.1). The position of a point on the line can be described by a single "number": its distance from the origin, shown as x_0 in Figure 2.1. This is the familiar idea of a number-line from high school algebra. Since you need a single number to describe the position of a point on a line, you say, figuratively, that a number-line has a *dimension* of 1. You should be aware of the fact that mathematician sometimes think of spaces, numbers, distance, and dimensions in terms more abstract than their everyday use.

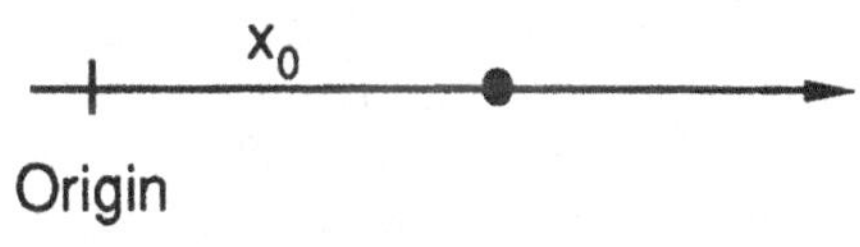

Figure 2.1.

Denote the distance of the point from the origin by x_0, and let us apply the following *transformation* which moves our point to a new location on the line:

$$x_1 = a\,x_0 \qquad (1)$$

where a is a *parameter* (a real number) and x_1 is the new distance from the origin. The next position of the point (that is x_1) depends on the value of a. We have the following possibilities:

- if $a < -1$, x_1 will move to the left of the origin, and will be further away from the origin than x_0 was;
- if $a = -1$, x_1 will move to the left of the origin, and will be at the same distance as x_0 was;

- if $-1 < a < 0$, x_1 will move to the left of the origin and will be closer to it than x_0 was;
- if $a = 0$, x_1 will be at the origin;
- if $0 < a < 1$, x_1 will be to the right of and closer to the origin than x_0;
- if $a = 1$, the position of the point remains unchanged, that is $x_1 = x_0$; finally,
- if $a > 1$, x_1 will move further away from the origin and to the right.

We can now generalize our transformation to more than one step, and write equation (1) as:

$$x_{n+1} = a\, x_n \;,\; x_0 \text{ given.} \tag{2}$$

Equation (2) allows us to calculate the new position of each point based on its previous position, given that the *initial* position, x_0, is known. The transformation (2) is also called a ***map***, or an ***iteration***, and it is linear in a sense that no powers of x (other than 0 and 1, of course) are involved.

Next, suppose we want to move the point around on the line as in equation (2), but after calculating its new position we want to "nudge" it by a constant distant. We call this nudging *translation*, and we achieve it by adding a free parameter to equation (2) to get

$$x_{n+1} = a\, x_n + b \;,\; x_0 \text{ given.} \tag{3}$$

where b is a real number. For example, if $b = 1$, x_{n+1} would be moved one unit to the right.

To describe the position of a point on a plane, we need two coordinates, x and y, as shown in Figure 2.2. The distances x_0 and y_0 are also called the x,y coordinates of the point, and by convention are written as (x,y). Thus, if the position of the point shown in Figure 2.2 is, for example, 10 units along the x axis and 15 units along the y axis we write (10,15) for its coordinates. The position of a point is then described by its distance from the origin along the x-axis, denoted as x_0 in Figure 2.2, and along the y-axis, denotes as y_0 in Figure 2.2. We now have two independent directions in our space (plane), and we conclude that the dimensionality of a plane is 2. Next, let us see what we can do with transformations of points on the plane.

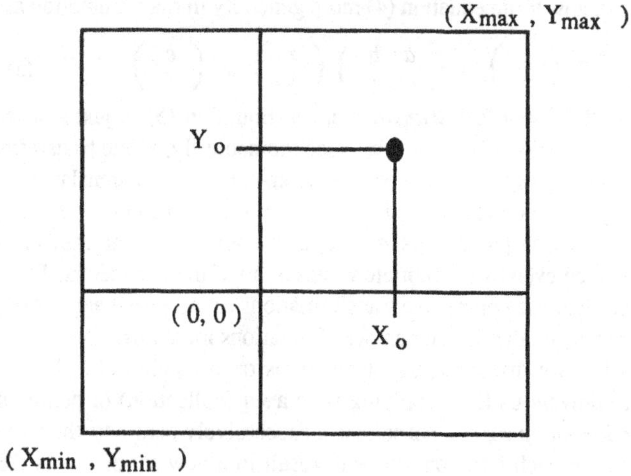

Figure 2.2.

Similar to equation (3), we write

$$\left.\begin{array}{ll} x_{n+1} = a\,x_n + b\,y_n + e & x_0 \text{ given} \\ y_{n+1} = c\,x_n + d\,y_n + f & y_0 \text{ given} \end{array}\right\} \qquad (4)$$

where the parameters *a*, *b*, *c*, *d*, *e*, and *f* are all real numbers. We can now move the point (shown in Figure 2.2) around on the plane to wherever we desire by playing with the parameter values. For example, if we set all of the parameters to zero, one iteration suffices to move any point in the plane to the origin, where the point will stay for as many iterations as we apply (that is for any *n*). We can also move to the origin "slowly" by choosing *a*, *b*, *c*, and *d* to all be strictly between -1 and 1, and setting $e = f = 0$. We can also translate our original point to any place we wish on the plane by choosing $-1 < a, b, c,$ and $d < 1$ and setting *e* and *f* to the appropriate values (found by solving $x^* = ax^* + by^* + e$ and $y^* = cx^* + dy^* + f$ with (x^*, y^*) being our final point, often called the fixed point) . Note also that—with an appropriate choice of the parameters *a*, *b*, *c*, and *d*—we can move the point (*x*,*y*) to the ***fixed point*** rapidly or slowly. Finally, note that if we choose the parameters as described here, no matter where we start on the plane, we will wind up at the same point (provided that we apply the iteration (4) enough times).

We can now write equation (4) more generally in matrix notation as

$$W\begin{pmatrix} x \\ y \end{pmatrix} = \begin{pmatrix} a & b \\ c & d \end{pmatrix}\begin{pmatrix} x \\ y \end{pmatrix} + \begin{pmatrix} e \\ f \end{pmatrix} \qquad (5)$$

where W is an *affine transformation*. Equation (5) is just a short hand notation for equation (4). As you will see momentarily, affine transformations preserve images in space, but in general contract or expand, translate and rotate these images. A contraction can be achieved, for example, by setting the absolute values of parameters a, b, c, and d all < 1. The translation and rotation are achieved by appropriate values of all of the parameters. To produce fractals we shall use contractive transformations (in a sense that pairs of points get closer to each other as the number of iterations increases).

Thus far, we discussed transformations on a single point. But there is nothing to prevent us from applying W to a *set* (collection) of points. If W is a *contraction mapping* in a sense that it successively contracts the size of the initial set after each iteration, this will result in a new set of points closer to each other. Applying W many times, we should eventually obtain a single point, called the *fixed-point* of the affine transformation W. To see this, consider the following 2 examples.

Start with a set of points which form the sides of a square, and denote that square by S_0 (Figure 2.3). Now to each point on the edge of the square (keep in mind that the interior of the square is not part of our initial set; the edges are) apply the transformation

$$W\begin{pmatrix} x \\ y \end{pmatrix} = \begin{pmatrix} 0.75 & 0.00 \\ 0.00 & 0.75 \end{pmatrix}\begin{pmatrix} x \\ y \end{pmatrix} + \begin{pmatrix} 0.00 \\ 0.00 \end{pmatrix} \qquad (6)$$

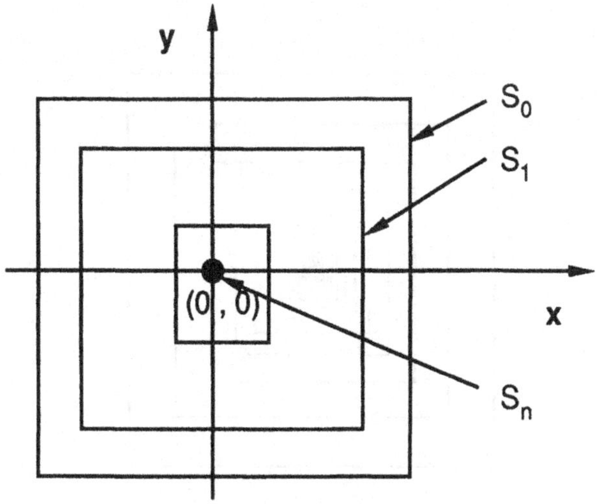

Figure 2.3

In other words, take each point on the square S_0 and move it closer to the origin by 1/4 of its distance from the origin. If you repeat this process many times, you will eventually get to the origin with your original set, S_0. This final set is shown as S_n in Figure 2.3. Figure 2.4, which gives you an indication of things to come, shows the actual sequence of iterations as achieved by using *Fractal Attraction*. We do not produce the final point, since the screen resolution essentially renders the picture black after about 10 iterations. But if you were to zoom in the middle, what seems like a black band is actually many squares, close to each other.

It should be clear by now that if you start with a circle, with the affine transformation (5), you will end up with a point at the origin (Figure 2.5). In fact, you can start with a picture of Mount Everest if you wish, and end up with a single point at the origin (given a large enough number of iterations).

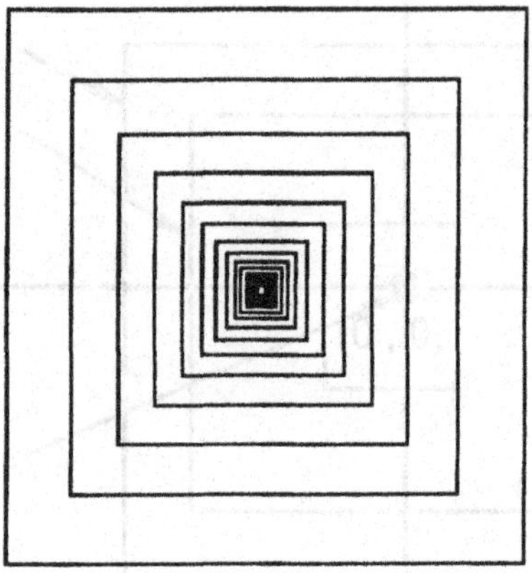

Figure 2.4.

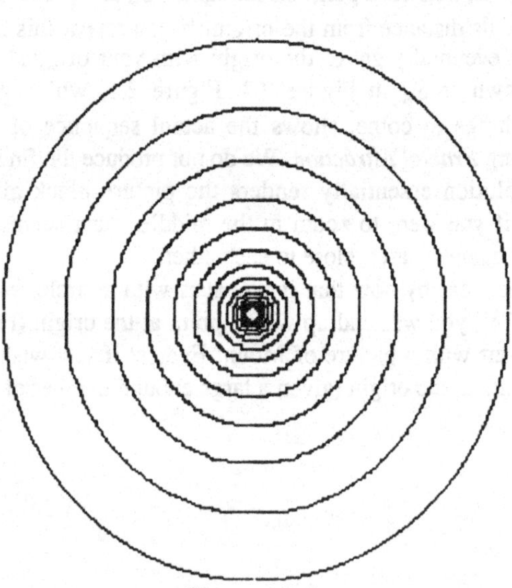

Figure 2.5.

§2.2. *Iterated Function Systems (IFS)*

Just as we generalized the notion of applying an affine transformation to a set (where the set is the actual image we see), we can generalize and apply a set of affine transformations to a set of initial points, which we call the ***initial set***. We thus have the following definitions.

Definition 1. An *IFS* is a collection of contractive affine transformations.

It so happens that iterations of IFS on an initial set result in a unique image, regardless of the initial image. We thus say that

Definition 2. The unique image that is produced by iterations of an IFS is a *fractal attractor* of the given IFS.

The remarkable fact is that no matter what our initial image is, after many iterations of the IFS we will always wind up with the same image: the attractor.

Let us analyze the following example, where 3 affine transformations are used:

$$W_1\begin{pmatrix} x \\ y \end{pmatrix} = \begin{pmatrix} 0.50 & 0.00 \\ 0.00 & 0.50 \end{pmatrix}\begin{pmatrix} x \\ y \end{pmatrix} + \begin{pmatrix} -5.00 \\ -5.00 \end{pmatrix} \quad (7)$$

$$W_2\begin{pmatrix} x \\ y \end{pmatrix} = \begin{pmatrix} 0.50 & 0.00 \\ 0.00 & 0.50 \end{pmatrix}\begin{pmatrix} x \\ y \end{pmatrix} + \begin{pmatrix} 5.00 \\ -5.00 \end{pmatrix} \quad (8)$$

$$W_3\begin{pmatrix} x \\ y \end{pmatrix} = \begin{pmatrix} 0.50 & 0.00 \\ 0.00 & 0.50 \end{pmatrix}\begin{pmatrix} x \\ y \end{pmatrix} + \begin{pmatrix} 5.00 \\ 5.00 \end{pmatrix} \quad (9)$$

If you start with a square, after applying W_1 in equation (7) to each point on the square once, you get a new square that is squeezed half way in both the x and y directions, and is translated -5 units in the x direction and -5 units of distance in the y direction. Equation (8) results in a similar transformation, except that now the original square is translated 5 units in the x direction and -5 units in the y direction. Finally, equation (9) takes the original square, squeezes it by half in each direction, and translates it 5 units of distance to the right and 5 units of distance up. Thus, if we start with an original square whose corners sit in (-10,-10), (10,-10), (10,10), and (-10,10), and apply each transformation once, we get 3 squares, as shown in Figure 2.6.

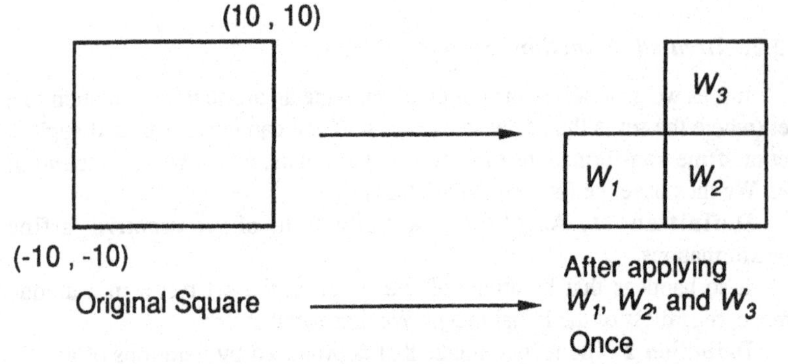

Figure 2.6.

After applying the iteration again, we get from each old square 3 new squares, each squeezed by half in each direction and translated according to the free parameters in equations (7) - (9); we thus get 9 new squares, and so on. After 9 more iterations we get the picture in Figure 2.7.

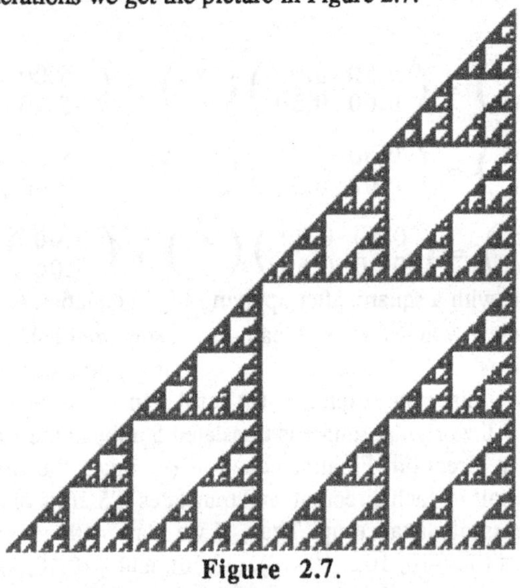

Figure 2.7.

Figure 2.7 is called the *Sierpinski triangle*. Now what do you think you will see if you take any region in Figure 2.7 and blow it up? You are right, you will see sets of 3 squares stacked as shown in the right side of Figure 2.6. The fact that zooming results in the same pattern as the original image (as shown in Figure 2.7) is one of the most distinguishing characteristics of images produced by IFS. Such *self similar* images are called *fractals*. You can now see how easy it is to produce *fractal attractors* using IFS.

If you are not convinced that—given enough iterations of the IFS (7) to (9)—any initial image will end up as a Sierpinski triangle (which is the fractal attractor of the IFS in equations (7) - (9)), look at the sequence of Figures 2.8 through 2.11 below. In Section §2 we shall explain in detail how this sequence of figures was produced with *Fractal Attraction*. Briefly, we first produce the fractal attractor for the following IFS code:

$$W_1\begin{pmatrix} x \\ y \end{pmatrix} = \begin{pmatrix} 0.000 & 0.000 \\ 0.000 & 0.172 \end{pmatrix}\begin{pmatrix} x \\ y \end{pmatrix} + \begin{pmatrix} 0.496 \\ -0.091 \end{pmatrix} \quad (10)$$

$$W_2\begin{pmatrix} x \\ y \end{pmatrix} = \begin{pmatrix} 0.076 & 0.312 \\ -0.257 & 0.204 \end{pmatrix}\begin{pmatrix} x \\ y \end{pmatrix} + \begin{pmatrix} 0.494 \\ 0.133 \end{pmatrix} \quad (11)$$

$$W_3\begin{pmatrix} x \\ y \end{pmatrix} = \begin{pmatrix} 0.821 & -0.028 \\ 0.030 & 0.845 \end{pmatrix}\begin{pmatrix} x \\ y \end{pmatrix} + \begin{pmatrix} 0.088 \\ 0.176 \end{pmatrix} \quad (12)$$

$$W_4\begin{pmatrix} x \\ y \end{pmatrix} = \begin{pmatrix} -0.024 & -0.356 \\ -0.323 & 0.074 \end{pmatrix}\begin{pmatrix} x \\ y \end{pmatrix} + \begin{pmatrix} 0..470 \\ 0.260 \end{pmatrix} \quad (13)$$

which results in a fern (Figure 2.8). After 1 iteration of the Sierpinski triangle IFS (as given in equations 7 - 9) on the initial image in Figure 2.8 we get 3 ferns, as shown in Figure 2.9. An additional iteration of the IFS in equations (7) - (9) gives us 9 ferns (Figure 2.10), and so on. Note the distinction between transformations and iterations. A transformation is a particular equation, such as (10). When we say that we iterate, we mean that we calculate a new position for each point in the current image. This calculation is made for each of the transformations, which together form the IFS. After 10 iterations we get the Sierpinski triangle (Figure 2.11) which is now made of many ferns (in fact as many as 3^{10} ferns) arranged in triangles of 3. If you blow up the picture you will be able to see it.

In summary, we have seen that an IFS describes a unique fractal attractor. Repeated applications (iterations) of the IFS to any initial image (set of points) will result in that unique attractor. So now we have a method to store images as a set of equations! The problem is then how to derive an IFS for a given image. More on that later.

Figure 2.8. The initial set.

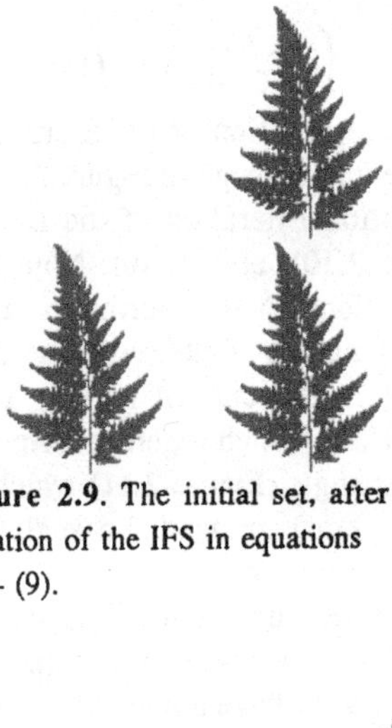

Figure 2.9. The initial set, after one iteration of the IFS in equations (7) - (9).

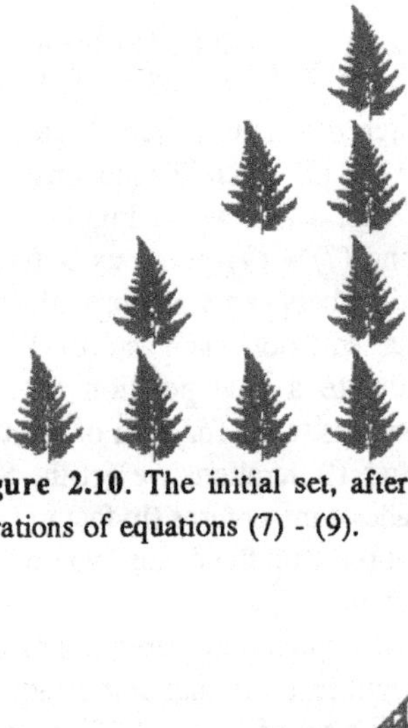

Figure 2.10. The initial set, after 2 iterations of equations (7) - (9).

The initial set, after 5 iterations of the IFS in equations (7) - (9).

Figure 2.11. The initial set, after 10 iterations of the IFS in equations (7) - (9).

§2.3. *How to Design an Attractor*

In this section, we shall discuss how to design fractal images. What we will show you is how to create an IFS which, when iterated, creates an image of interest. You should be aware that there is also a way to take an existing image and create an IFS which will give you the image when you apply it to any initial image (set of points). The fact that you can actually do that is based on the so-called collage theorem (Barnsley 1988). In the process of explaining how to design an image, we shall also explain how *Fractal Attraction* works. When you are ready to design your own attractor we suggest you see section §4, **Using the Design Window** and §5, **Applying the Collage Theorem**. A more detailed explanation of the menu commands is given in section §7.2. To best benefit from this section, follow the instructions as outlined below.

So let us start. Double-click on the *Fractal Attraction* icon. Three windows, titled **Fractal**, **Design**, and **Untitled** are created. *Fractal Attraction* draws the image in the **Fractal** window. What each affine transformation in the IFS does after a single iteration is shown in the **Design** window. For this section, work with the **Untitled** widow, which contains the actual IFS code. You name the IFS code file (some are supplied with *Fractal Attraction* in a folder called IFS Examples). In this window you enter the parameters of your IFS.

Suppose you want to design a "fractal" line. You need to use 2 contractive affine transformations to produce a line (one transformation will always produce a point). The parameters *a*, *b*, *c*, *d*, *e*, and *f* are given in a table with appropriate cells in the **Untitled** window. Since a new **code** window defaults to 1 transformation, you do need to change the number of transformations to 2. *Fractal Attraction* will let you specify up to fifteen transformations. Each row corresponds to a single transformation. Now enter the following values in the table:

2 Number of Transformations [1 to 15]

	Transformation								Probability
1.	x_{new}	=	0.500	0.000	x_{old}	+	0.500		
	y_{new}		0.000	0.500	y_{old}		0.500		0.500
2.	x_{new}	=	0.500	0.000	x_{old}	+	0.000		
	y_{new}		0.000	0.500	y_{old}		0.000		0.500

Notice that there is one more column, titled ***probability***. Ignore it for now. What you have actually done is create the following IFS:

$$W_1\begin{pmatrix} x \\ y \end{pmatrix} = \begin{pmatrix} 0.50 & 0.00 \\ 0.00 & 0.50 \end{pmatrix}\begin{pmatrix} x \\ y \end{pmatrix} + \begin{pmatrix} 0.50 \\ 0.50 \end{pmatrix} \quad (14)$$

$$W_2\begin{pmatrix} x \\ y \end{pmatrix} = \begin{pmatrix} 0.50 & 0.00 \\ 0.00 & 0.50 \end{pmatrix}\begin{pmatrix} x \\ y \end{pmatrix} + \begin{pmatrix} 0.00 \\ 0.00 \end{pmatrix} \quad (15)$$

When you scroll to the bottom of the edit window you will see a figure called Fractal Space. It is here that you specify the coordinates of your plane (space). *Fractal Attraction* defaults to *(-0.1, -0.1)* for the lower left corner of the plane and *(1.1,1.1)* for the upper right corner. These are the bounds used by the **Fractal** and **Design** windows.

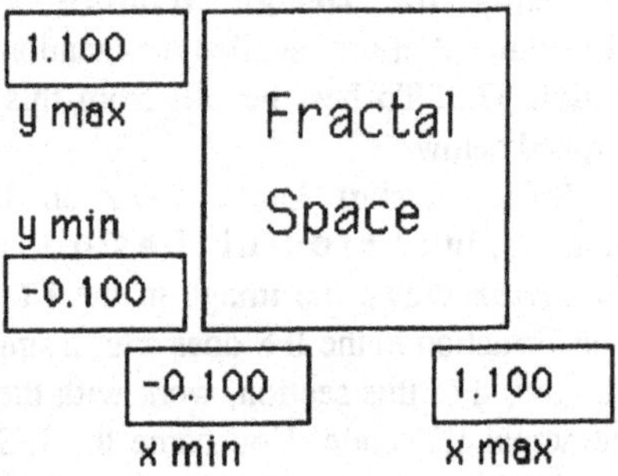

Under the **Code** menu you will find the menu item **Calc. X & Y Scale**. The effect of **Calc. X & Y Scale** is to automatically scale the minimum and maximum values of *x* and *y* in your drawing region to include the majority of the image. You can, if you wish, scale the picture yourself. You may want to scale your picture if you are interested in zooming in on a particular region in your image. Use the default values for now. They result in 2 slightly overlapping squares in the **Design** window, as shown in Figure 2.12.

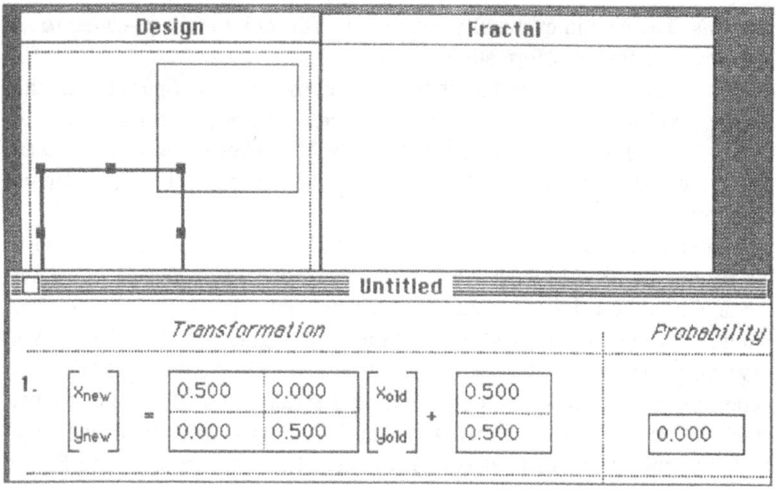

Figure 2.12.

Observe Figure 2.12 carefully. Can you figure out what happens? Here is the answer: the **Design** window shows a single iteration of the IFS code on an initial outline of a square (dotted lines). The first transformation produces the upper square on the right, and the second the lower square on the left. To find out which transformation does what you can use either the **Design** or the **Fractal** window.

Use the **Design** window first. Start by arranging your screen as shown in Figure 2.12. This way you will be able to see the **Design** window as you work in the **code** (**Untitled**) window. You can click anywhere on any cell in the IFS code table (do it now). As you can see, the square which pertains to the transformation (whose parameters are given in the line) is highlighted. Switch back and forth clicking the first and the second rows and observe the squares that become highlighted in the **Design** window. Change some of the numbers (parameters) in the first row. See if you can determine the effect of changing parameters *a* through *f*.

The second way to see what a particular transformation does is as follows. Click on the **Fractal** window (this will make that window active). Now choose your initial set by going to the **Add Figure** submenu under the **Fractal** menu, and choosing (for example) **Circle**. You will get a circle in your window. The points on this circle define your initial set. Next, go to the **Fractal** menu and choose **Deterministic**. Next, choose **Trans 1** from the **Window** menu (you can use the ⌘-1 key combination, that is, press both the ⌘ and 1 at the same time) to iterate the first affine transformation once. Do it now and observe what happens: your circle is now smaller; it also moved up and to the right. Next repeat ⌘-1 a few times and observe what happens. You should eventually get a point. This point is the so-called *fixed-point* of the affine transformation.

To see what the second transformation does, erase the figure by clicking on the **Edit** menu and choosing **Clear**. Now choose, from the **Fractal** menu, an initial set, the circle again, and type ⌘-2 a few times to see what the second transformation does. Find where the fixed-point of our second transformation is.

Next, make sure that the **Fractal** window is active by clicking on it. Verify that the **Deterministic** algorithm is checked under the **Fractal** menu. Go to the **Edit** menu and click **Clear**. Put a square in the **Fractal** window by clicking **Square** under the **Add Figure** submenu. What you have done then is initiate your initial set to a square and chose to apply the deterministic algorithm to it (the **Random** algorithm will be explained later). **Do One** under the **Fractal** menu 5 times (you can alternatively type ⌘-D to achieve the same effect). Did you get the picture shown in Figure 2.13?

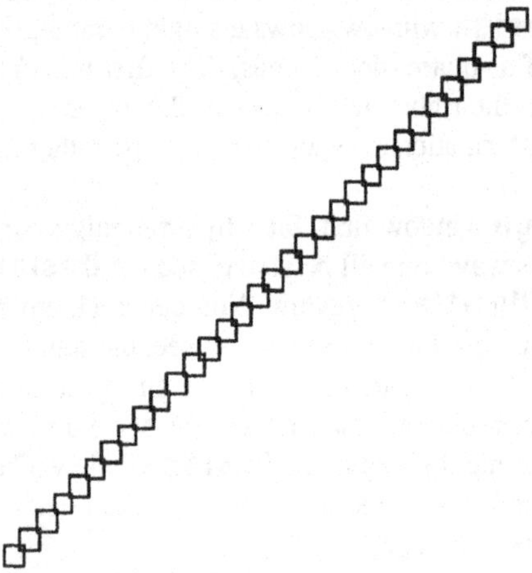

Figure 2.13.

If you want to trace the history of iterations (without erasing the result of the previous iterations) follow these steps: Add an identity transformation.

$$W_3\begin{pmatrix} x \\ y \end{pmatrix} = \begin{pmatrix} 1.00 & 0.00 \\ 0.00 & 1.00 \end{pmatrix}\begin{pmatrix} x \\ y \end{pmatrix} + \begin{pmatrix} 0.00 \\ 0.00 \end{pmatrix}.$$

This transformation preserves the original figure. Make the edit window the active window. Change the number of transformations to 3. This will add a third row in which you enter 1 in the *a* and *d* cells; the rest of the third row parameters should be set to zero. Erase your figure again (by making sure that the **Fractal** window is active and then by choosing **Clear**). Then initiate your figure to a square (by choosing **Square** from the submenu). Make sure that **Deterministic** is checked and choose **Run** from the **Fractal** menu. In response to the dialog box enter 5 for the number of iterations. Did you get the picture in Figure 2.14?

Let us go back to the **Design** window for a moment. Suppose that you do not like that little overlap in the squares that you get from your IFS code. We now try to make the squares touch at the tip. Play around with the parameters *e* and *f* until you get the squares in the **Design** window just touching. You may have to change *x*-minimum, *x*-maximum, *y*-minimum, and *y*-maximum to achieve the desired result. Once you get those squares to just touch, initiate your figure to a circle and do 5 iterations. Did you get the picture shown in Figure 2.15?

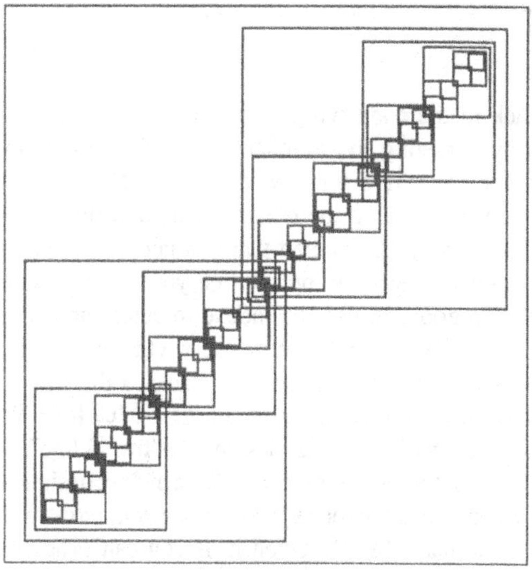

Figure 2.14.

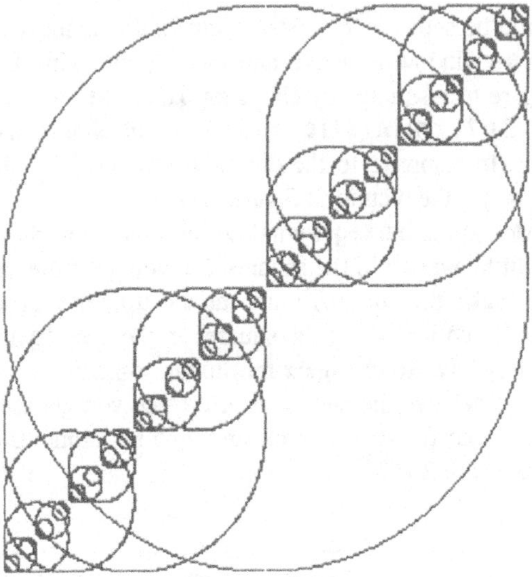

Figure 2.15.

Now go back to the **code** window and remove the third (the identity) transformation by changing the number of transformations to 2. Make the **Fractal** window the active window again and clear the figure. Choose **Circle** again and try running the deterministic algorithm another 5 times. Play around a little, convince yourself that you get a line from any shape you start, and that when you zoom in on the line you can see the original shape you began with. To zoom in use the mouse to select part of the **Fractal** window then choose **Zoom In** under the **Fractal** menu. This will change the coordinate system. If you want to get back to the original coordinate system make the edit window active and choose **Calc. X & Y Scale**. This concludes our discussion of the fractal design of a "fractal line".

We shall demonstrate one more example, which is designed to show you what happens when the parameters *b* and *c* are not zero. In the folder IFS Codes there is a file named Crab. Open it. If you can't (because the **Open** command under the **File** menu is dimmed) it's because you haven't closed your current IFS code file. Remember, the current version of *Fractal Attraction* deals with one IFS code file at a time. So close the current IFS code file (save it with a different name if you wish by choosing **Save As...** from the **File** menu) before opening a new IFS code file. Now that Crab is open save it under a different name, say NewCrab, and play around with the parameter values *a*,*b*,*c*, and *d* in the first transformation (the first row of the table) in **NewCrab** (this way you will not change the original Crab file). Observe what happens: the parameters *a* and *d* shrink the original square (you observe these effects on the **Design** window; so make it visible and then change parameters in the NewCrab window). The parameters *b* and *c* rotate the square in the appropriate directions. Now go back to the original **crab** file and try to produce Figure 2.16.

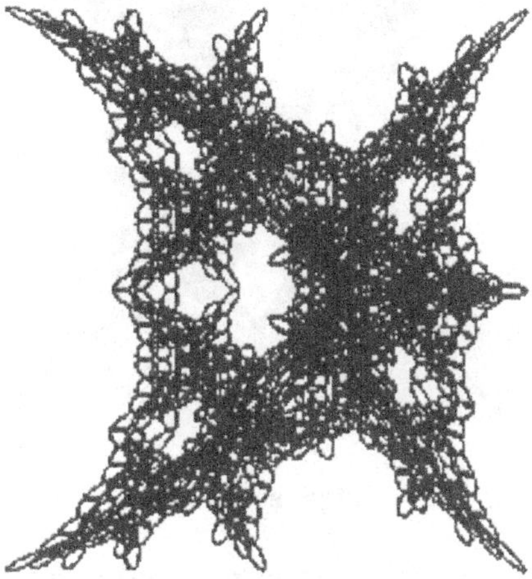

Figure 2.16.

Here is how we did it: we set the starting figure to a circle and did 5 **Deterministic** iterations. Our NewCrab, with the transformations:

$$W_1\begin{pmatrix} x \\ y \end{pmatrix} = \begin{pmatrix} 0.50 & 0.20 \\ 0.20 & 0.50 \end{pmatrix}\begin{pmatrix} x \\ y \end{pmatrix} + \begin{pmatrix} 0.00 \\ 0.00 \end{pmatrix} \quad (16)$$

$$W_2\begin{pmatrix} x \\ y \end{pmatrix} = \begin{pmatrix} -0.50 & -0.20 \\ 0.20 & 0.50 \end{pmatrix}\begin{pmatrix} x \\ y \end{pmatrix} + \begin{pmatrix} 1.00 \\ 0.00 \end{pmatrix} \quad (17)$$

$$W_3\begin{pmatrix} x \\ y \end{pmatrix} = \begin{pmatrix} 0.50 & 0.20 \\ -0.20 & -0.50 \end{pmatrix}\begin{pmatrix} x \\ y \end{pmatrix} + \begin{pmatrix} 0.00 \\ 1.00 \end{pmatrix} \quad (18)$$

$$W_4\begin{pmatrix} x \\ y \end{pmatrix} = \begin{pmatrix} -0.50 & -0.20 \\ -0.20 & -0.50 \end{pmatrix}\begin{pmatrix} x \\ y \end{pmatrix} + \begin{pmatrix} 1.00 \\ 1.00 \end{pmatrix} \quad (19)$$

produced Figure 2.17 after 5 iterations (choosing **Do One** 5 times) from an initial square.

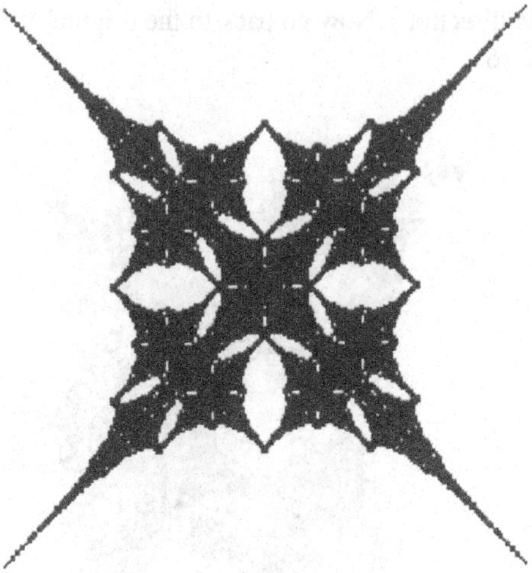

Figure 2.17

Note that one deterministic iteration beginning with a circle with the IFS code (16) - (19) results in Figure 2.18, and 4 of them result in Figure 2.19.

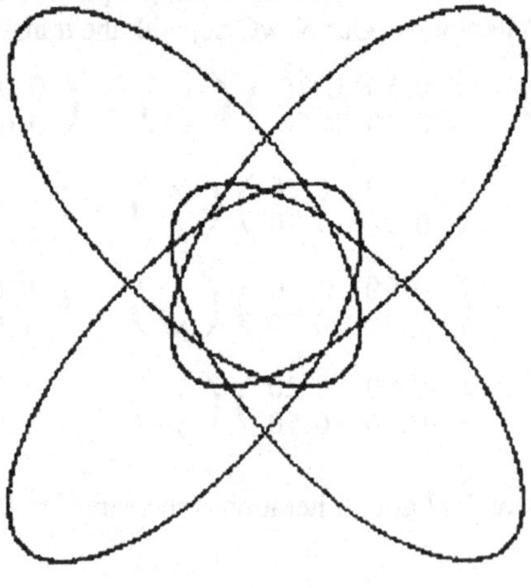

Figure 2.18

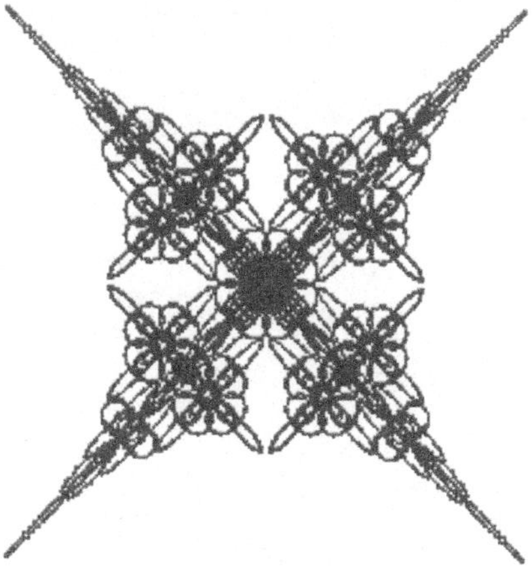

Figure 2.19.

Try to eliminate some of the transformations in your NewCrab IFS code; see what happens. Eliminating a transformation is easily done from the **design** window. Click on the transformation you would like to eliminate in the **Design** window, it will become selected, then choose **Clear** under the **edit** menu. Have fun, and if you discover a beautiful attractor, let us know by picture or by IFS code.

§3. The Random Algorithm

During the last section (Mathematical Background) the deterministic algorithm was used to generate the fractal images. This section will explain an alternative algorithm, the random algorithm, and how to use it. It has the advantage that it is usually faster than the deterministic algorithm. It also allows interesting coloring schemes for the attractor which are covered in the color rendering section.

For an introduction to the random algorithm play the "Chaos Game." You will need a ruler, a pencil and a die. Make a copy of the game sheet as shown below .

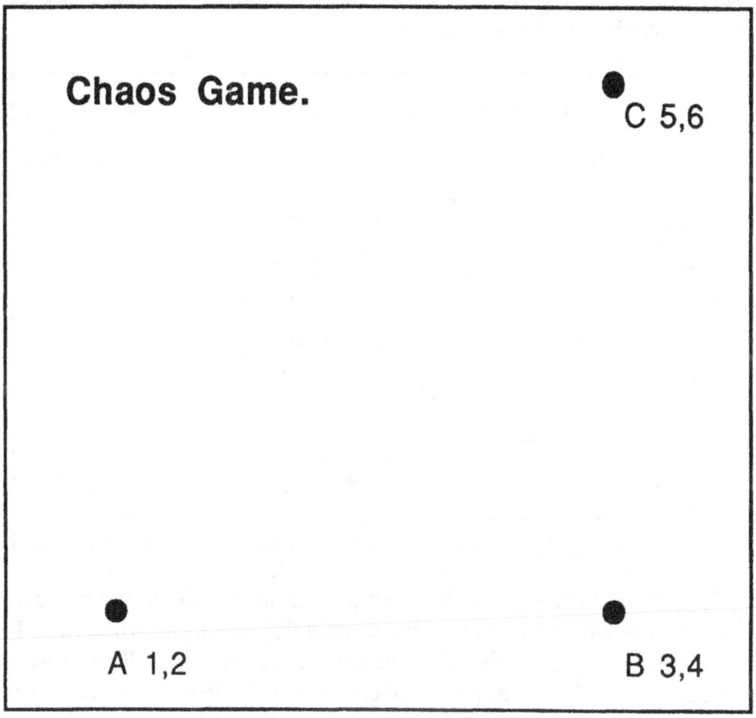

Chaos Game Sheet

Points **A**, **B**, and **C** are called fixed points. To play the game roll the die to determine which of the fixed point will be the starting point. If you roll a one or a two choose **A** as your initial point, a three or a four start at **B**, otherwise start at **C.** For the purpose of learning the game pretend the die was a one or two, therefore your first point is **A.**

A move to the next point involves moving towards one of the three fixed points. Roll the die. Suppose you get 3, this means a move towards **B.** The rules involve moving half the distance from the current point to the selected fixed point. In this case since the current point is **A**, measure and mark the point half way from **A** to **B**. This will be the new current point; label its position by the number 2. Roll the die again, perhaps getting a four. From point 2 measure half the distance to **B** again, mark this point and label it 3. Let's say we get a 5 on the next roll. Measure off half the distance from point 3 to **C** and mark it as point 4. Do this about 10,000 times (if you have nothing else to do). What figure did you get?

Now use *Fractal Attraction* to play the Chaos Game. Double-click on the *Fractal Attraction* icon. Close the **code** window and choose **Open** from the **File** menu. Open the file titled Sierpinski triangle. You should see the following numbers in the edit window.

3 Number of Transformations [1 to 9]

	Transformation					Probability
1.	$\begin{bmatrix} x_{new} \\ y_{new} \end{bmatrix}$ =	$\begin{bmatrix} 0.500 & 0.000 \\ 0.000 & 0.500 \end{bmatrix}$	$\begin{bmatrix} x_{old} \\ y_{old} \end{bmatrix}$	+	$\begin{bmatrix} 0.000 \\ 0.000 \end{bmatrix}$	0.330
2.	$\begin{bmatrix} x_{new} \\ y_{new} \end{bmatrix}$ =	$\begin{bmatrix} 0.500 & 0.000 \\ 0.000 & 0.500 \end{bmatrix}$	$\begin{bmatrix} x_{old} \\ y_{old} \end{bmatrix}$	+	$\begin{bmatrix} 0.500 \\ 0.000 \end{bmatrix}$	0.330
3.	$\begin{bmatrix} x_{new} \\ y_{new} \end{bmatrix}$ =	$\begin{bmatrix} 0.500 & 0.000 \\ 0.000 & 0.500 \end{bmatrix}$	$\begin{bmatrix} x_{old} \\ y_{old} \end{bmatrix}$	+	$\begin{bmatrix} 0.500 \\ 0.500 \end{bmatrix}$	0.340

The *probability* column corresponds to throwing the die. In this case each of the transformations (rows) are chosen (roughly) with equal probability. To run the program make the **Fractal** window active. Make sure **Random** is checked under the **Fractal** menu and choose **Do 10,000**. You should now see the same figure you got by hand. If you would like to see *Fractal Attraction* "play" the game in a little more detail choose **Trace** under the **Fractal** menu. This will allow you to follow the points as they are generated. Click the mouse to get out of the trace mode.

Do you understand the relationship of the transformations and the chaos game? Applying transformation one is the same as moving half the distance to point **A**. Transformation two is a move towards point **B** and three is a move half way to **C**. To see this a little more clearly bring the **Design** window to the front by clicking on it. Under the **Design** menu select **Show Fixed Points.** You will just barley be able to make out the fixed points on the outer edges of the small squares. These three points are the points **A, B,** and **C** from the chaos game. The computer picks a starting point [actually it starts at (0,0) and takes ten moves before plotting any points], generates a random number between 0 and 1 [rolling the die], and based on the random number it plots the next point by applying the selected transformation.

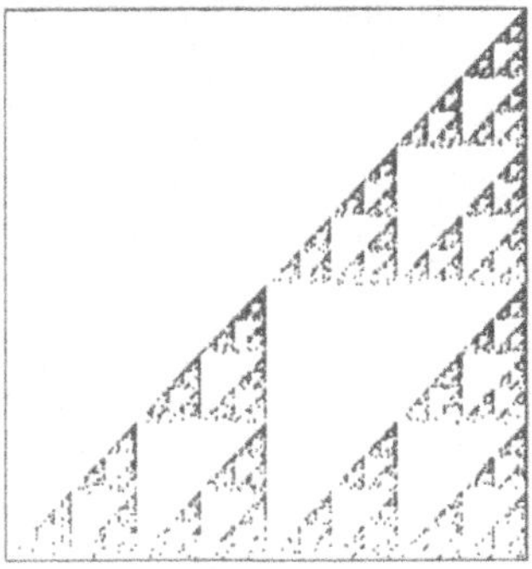

Fractal Attraction allows you to "fix" the die by changing the probabilities in the **code** window. Try entering 0.2 , 0.2 and 0.6. To run the algorithm the probabilities must sum to 1. Make the Fractal window active and clear the figure by choosing **Clear**. You should get a picture of the Sierpinski triangle that looks faded. Can you figure out why? By experimenting with the probabilities you can simulate various effects of shading the drawing.

Exercise: Try to change the Sierpinski IFS code to get this image of the triangle. Use the **Design** window to manipulate the transformations by pointing and clicking (see the next section for more information on using the design window).

There are other IFS codes that you can experiment with in the Fractal Examples folder. Have fun playing the Chaos Game.

§4. Using the Design Window

In the **Design** window the user can translate, rotate, shear and re-size transformations all by pointing, clicking and dragging. In fact, you will probably do all of your designing in this window, using the code window only to tweak the numbers. This section introduces these exciting features.

Start a session with *Fractal Attraction* by double-clicking on the program icon. An untitled document should be the top window. We will call it the **code** window. The single row of parameters corresponds to one transformation. Behind the code window are two more windows, **Design** and **Fractal**. Bring the **Design** window to the front either by clicking or choosing **Design** under the **Window** menu.

You should see two squares, a larger square with a dotted outline, and a smaller square (see figure 4.1). The larger square is the initial figure, or polygon, the smaller square is the result of applying transformation one (specified by the parameters in the code window) to the initial polygon. In this case the transformation consists of a contraction in both the x and y directions.

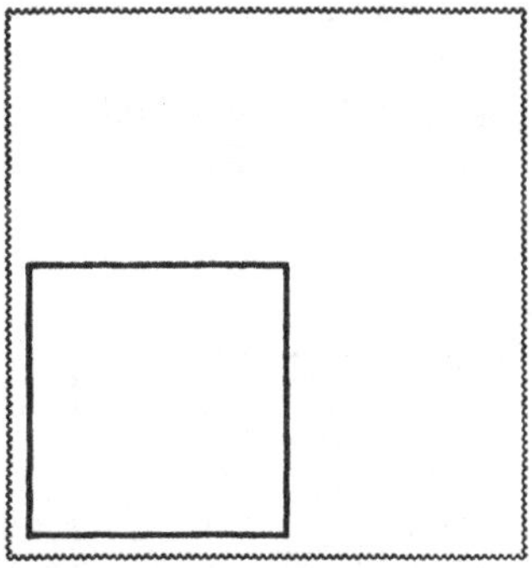

Figure 4.1 The **Design** window.

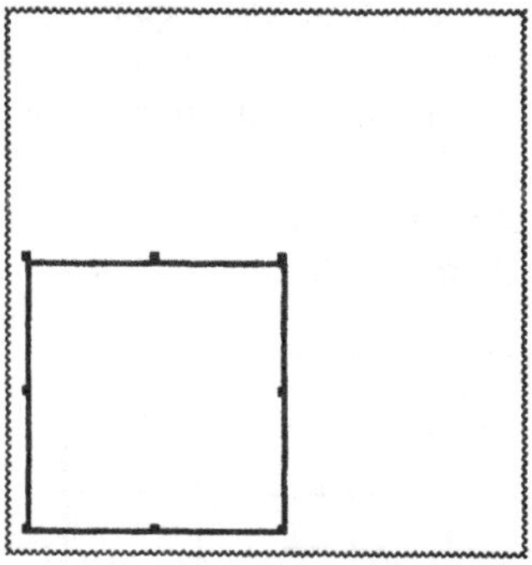

Figure 4.2 The selected transformation.

Selecting and Moving

Click the mouse inside the small square, it will now become selected. Eight small square or handles will appear on the boundary (see figure 4.2). Using these squares you will be able to rotate, shear, or re-size the square.

For now we just want to move it to a new position by dragging. Try it, hold the mouse down inside the small square and move to a new position. Bring the **code** window to the front and examine the parameters for transformation one. Do you see the effect of the move (better known as a linear translation)?

Arrange the **Design** window and the **code** window on the screen so that you can work in the **design** window yet still see transformation one in the **code** window. Try moving the small square to different positions and examine the transformation parameters.

Normally you will be working with several transformations. Most of the time you will be able to select a transformation merely by clicking. Occasionally the transformation you want will be covered by other transformations, in that case simply choose the transformation under the **Window** menu or click on the appropriate small rectangle on the right side of the design window.

Duplicating

A single transformation does not make a very interesting figure so add two more transformations. There are actually two ways to do this. One method is to use the **code** window and change the number of transformations. The second, slightly easier way is to duplicate an existing transformation in the design window.

To do this make sure the small square (transformation one) is still selected then choose **Duplicate Transformation** under the **Design** menu. You should see a second small square appear and it will now be selected. Note that a second row of numbers appeared in the **code** window. Choose duplicate one more time to get a third transformation. You should notice a third row of parameters in the **code** window. Note that duplicate does not duplicate the transformation exactly, it offsets (translates) the transformation to make it easier to work with (see figure 4.3).

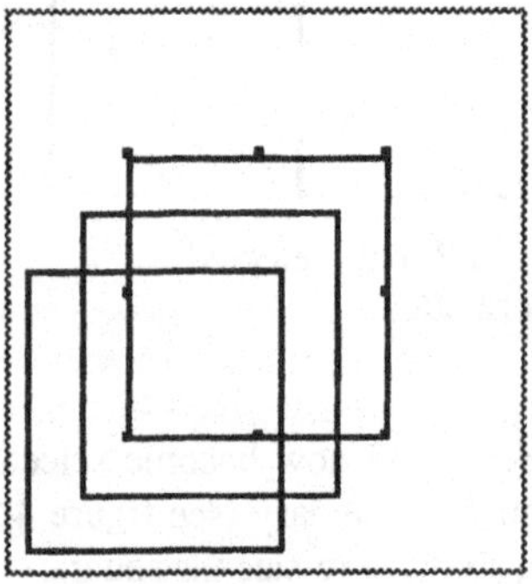

Figure 4.3
Duplicating transformations.

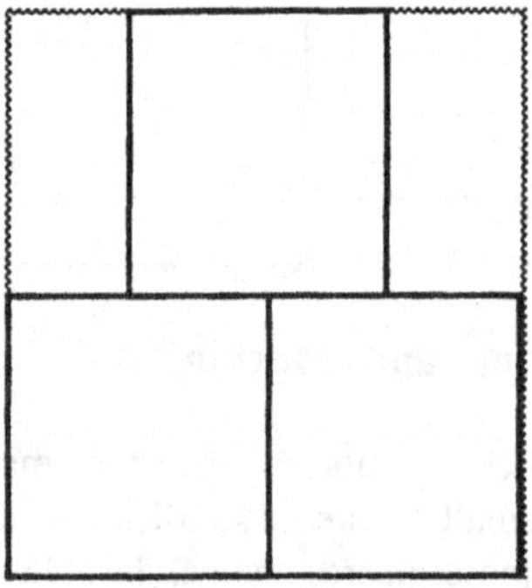

Figure 4.4
Moving transformations.

Clear

The **Clear** item under the **Edit** menu allows you to dispose of unwanted transformations. Simply select the transformation then choose **Clear.**

Sierpinski Triangle

The design window makes it easy to create the famous Sierpinski triangle. Using your moving skills arrange the three small squares so that they look similar to figure 4.4. Bring the **Fractal** window to the front by clicking or choosing it under the **Window** menu. Under **Fractal** menu choose **Do 10000**. Did you get an image similar to figure 4.5?

Figure 4.5
Sierpinski's Triangle, created in the **Fractal** window.

Re-sizing

By changing the size (or scaling factor) of transformation three you can get an interesting variation on Sierpinski's triangle. In figure 4.6 transformation three has been squeezed, the resulting fractal is figure 4.7.

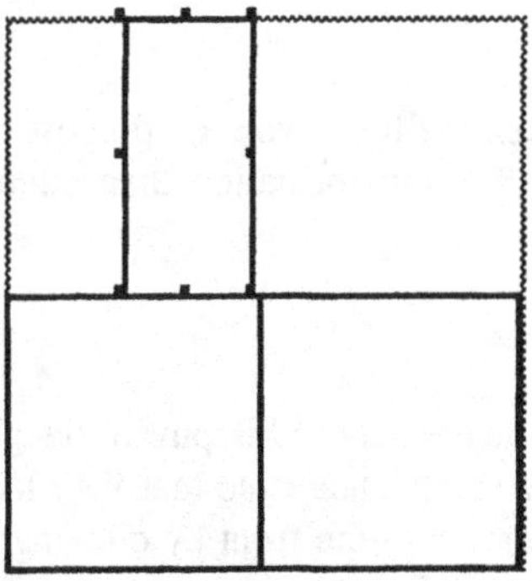

Figure 4.6
Transformation three has been squeezed.

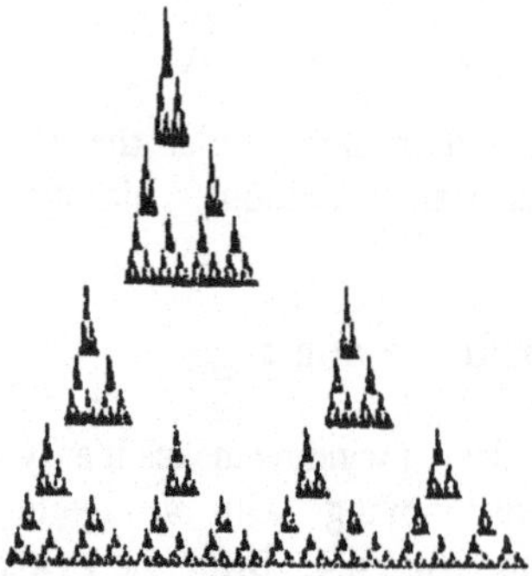

Figure 4.7
The result of the squeeze in **Fractal** window.

Clicking and dragging any of the eight small handles lets you resize the transformation. Experiment with this feature. To see the effect on the Sierpinski triangle make the **Fractal** window active and choose **Do 10000** again under the **Fractal** menu. You may want to clear the old figure by choosing **Clear** under the **Edit** menu.

If you hold the shift key as you click on one of the four corner handles the transformation will keep the same proportions while you are resizing it.

Rotating

Holding the option key down before selecting one of the four corner handles allows rotating the transformation. Figures 4.8 and 4.9 illustrate the effect of a forty-five degree rotation of transformation three. (Note that transformation three was expanded back to its original size before being rotated)

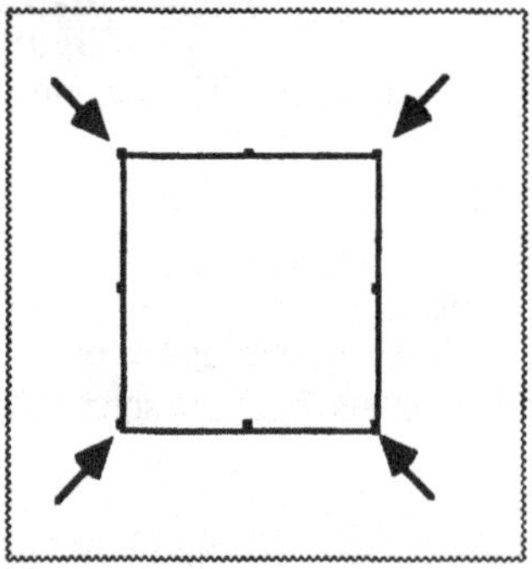

Hold the option key and drag one of the four corner handles to rotate.

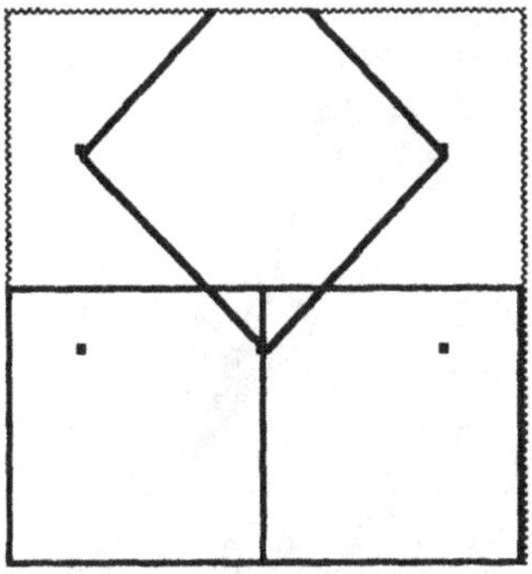

Figure 4.8
Transformation three has been rotated.

Figure 4.9
The resulting fractal.

Rotation is also available through the **Rotate** item under the **Design** menu.

Shearing

Shearing is accomplished by again holding the option key down but this time selecting one of the four middle handles. Figures 4.10 and 4.11 illustrate the effect of a shear (again note that transformation three was set back to the original position before shearing).

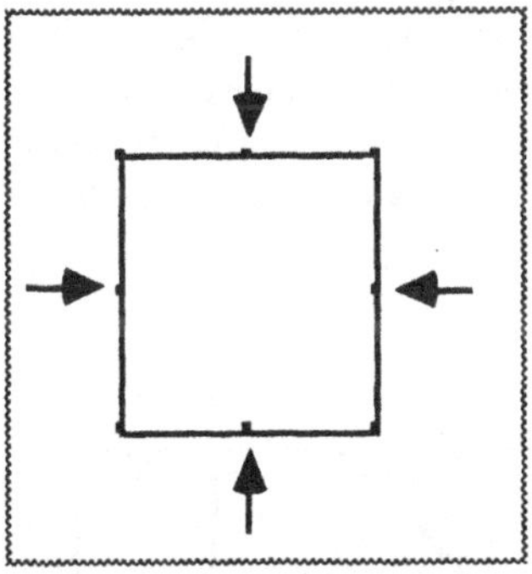

Use the middle handles to shear the transformation (while holding the option key).

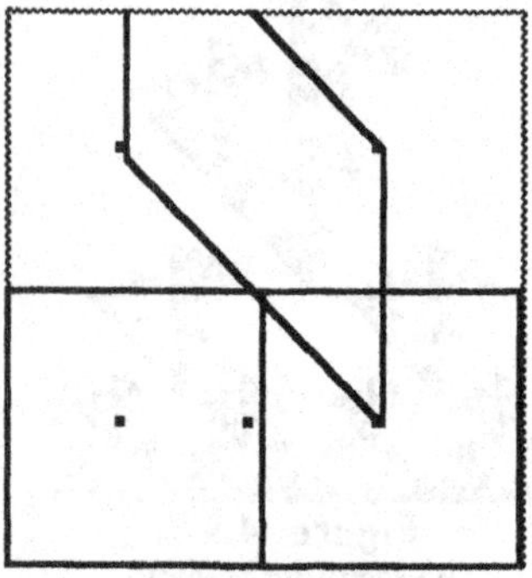

Figure 4.10
Transformation three has been sheared in the **Design** window.

Figure 4.11
The resulting fractal.

Reflections (Flips)

By applying the **Horizontal Flip** item under the **Design** menu to transformation three (see figure 4.10) we get figures 4.12 and 4.13. **Horizontal Flip** reflects all the x-coordinates across the vertical line that goes through the center of the transformation. **Vertical Flip** is similar but it reflects the y-coordinates across a horizontal line through the center.

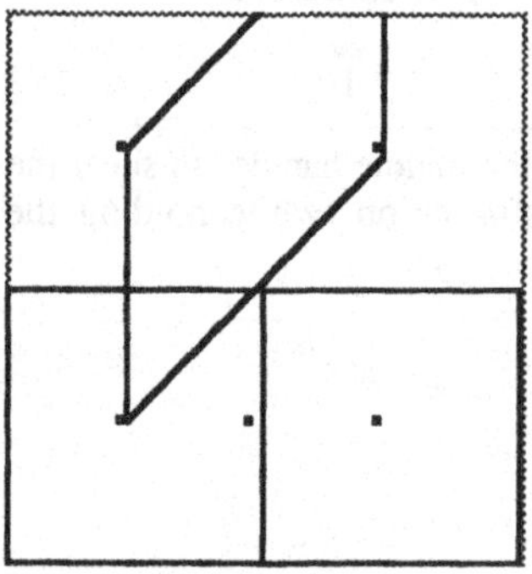

Figure 4.12
A horizontal flip of transformation three.

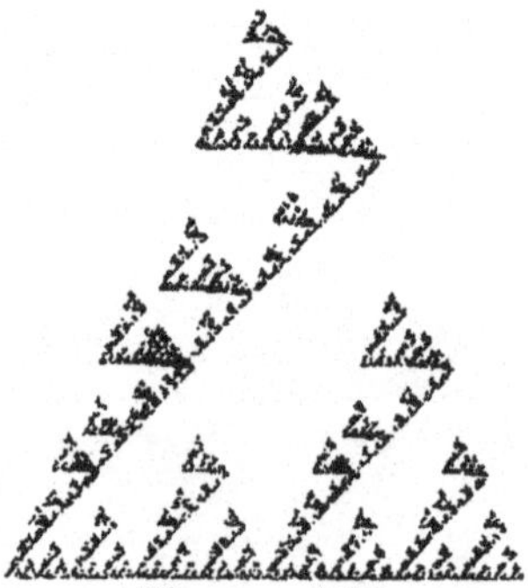

Figure 4.13

Initial Polygon

If desired *Fractal Attraction* will let you draw your own initial polygon instead of the default rectangle. Select **Draw Initial Polygon** under the **Design** menu. Click where you want to start and click for each corner of the figure (see figure 4.14). To complete the polygon click on the first point again (or double click from the last point). Figure 4.15 shows the effect of our original three transformations on the new initial polygon.

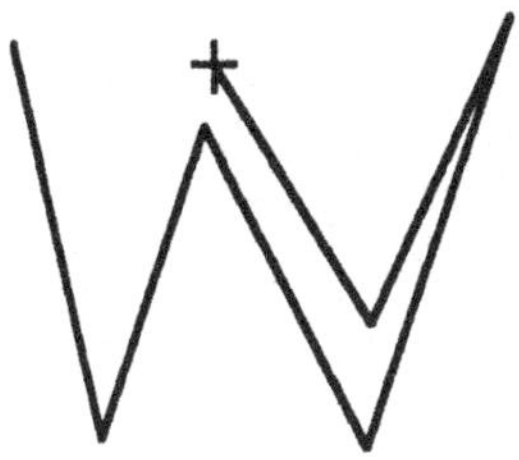

Figure 4.14
Entering an initial polygon in the **Design** window.

Figure 4.15
The new initial polygon and three transformations in the **Design** window.

This feature is useful if you want to try the Collage Theorem by Michael Barnsley (see figures 4.16 and 4.17). The next section gives an actual example of applying the collage theorem

Figure 4.16
A **W** as the new initial polygon and four transformations in the **Design** window.

Figure 4.17
The resulting fractal.

The initial figure can be saved by choosing **Save Initial Polygon** under the **Design** menu. Note that the figure will also be saved with your IFS code when you save the code window.

§5.Applying the Collage Theorem

Using Fractal Attraction to mimic nature (or one person's attempt at the collage theorem)

This example will illustrate using *Fractal Attraction* to mimic a maple leaf. To do this example completely from scratch you will need a scanner and a maple leaf. If you are not fortunate enough to have a scanner or live in a place like New York city where maple leafs are scarce, a MacPaint file of a leaf can be found on your original disk. The title of the file is "scanned_leaf". To use the file picture open it using a paint program, select the image using the smallest possible rectangle and copy to the clipboard. If you are working from scratch scan the maple leaf and copy to the clipboard. From the clipboard you will be able to paste it into *Fractal Attraction.*

Figure 5.1 Image in scanned_leaf file

The steps involved in building an IFS code to replicate the target image are listed below.

1. Paste target image (scanned maple leaf) into the **Design window.**
2. Outline the picture using the **Draw Initial Polygon** command.
3. Clear the picture (it is now outlined with a polygon) using the **Clear Picture** command.

4. Build an IFS code by covering the outline with reduced copies of itself.
5. Use the **Fractal** window to see how well the IFS code approximates the image! If the result is good save the IFS code, if bad return to step 4.

Step 1. Pasting in the target image.

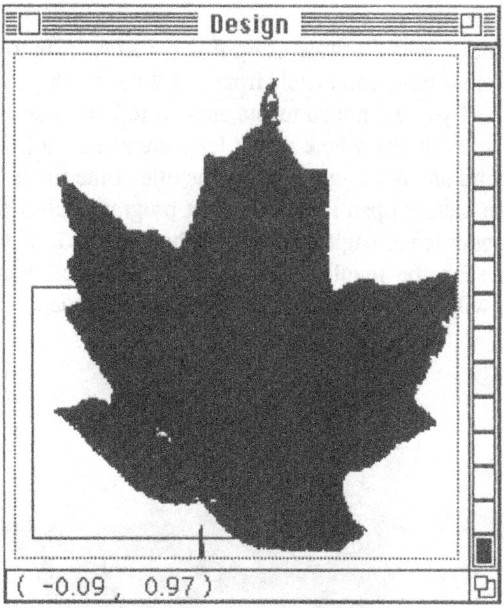

Figure 5.2 Leaf Picture pasted into **Design** window.

Start a new session with *Fractal Attraction.* This will give you a new untitled window for the code. Start working with the **Design** window, bring to the front by clicking or choosing **Design** under the **Window** menu.

Choose **Paste** under the **Edit** menu. A copy of the maple leaf from the clipboard will appear in the **Design** window (see figure 5.2). For now ignore the small square in the bottom left corner.

Resize the **design** window (see figure 5.3) so that the image is fully visible in the window. For this example the default coordinate system is fine (left. bottom is -0.1,-0.1, right, top is 1.1,1.1), if for some reason you want to change the coordinate system this would be the time to do it. The coordinates can be changed in the bottom of the code (untitled) window.

Figure 5.3 Resized Design Window.

Step 2. Outline the picture.

To get an outline of the leaf choose **Draw Initial Polygon** under the **Design** menu. You can now trace the leaf by clicking at points on the boundary. Click on the very top point of the leaf to get started. Now proceed around the leaf clicking at each point you want to include (see figure 5.4). Fractal Attraction is limited to about 30 points for the initial polygon so be frugal as you trace the leaf. The status bar at the bottom of the window will keep track of the number of points. To finish the polygon click on the first point again or double click for the last point.

Figure 5.4 Tracing the leaf image to form polygon.

Step 3.Clear the picture.

Once the leaf is outlined get rid of the leaf picture by choosing **Clear Picture** under the design menu. This will allow you to work with just the outline (referred to as the initial polygon) of the leaf. Notice that the small rectangle has been replaced by a half size copy of the leaf. Since the initial polygon is now the leaf outline it is the basis for the transformations. When you start a new session with *Fractal Attraction* you automatically get one transformation in the **code** window. The default initial polygon is a rectangle that outlines the window, as a result you get a half size copy of the rectangle. Since the initial polygon is now the leaf outline you get a half size copy of the leaf.

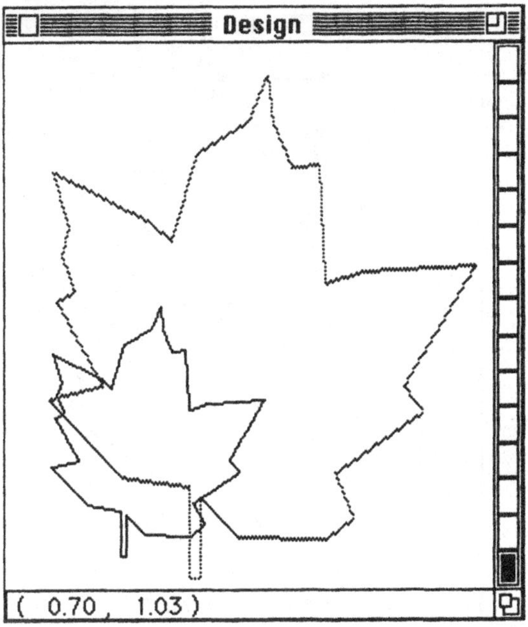

Figure 5.5 Outline of leaf and 1 transformation.

Step 4. Build an IFS code by covering the outline with reduced copies of itself.

This is the tricky part. The collage theorem by Michael Barnsley tells you that you can build an IFS code that will recreate the original set. The question is how to do it. One way is through a interactive program like *Fractal Attraction* which makes it easy for you to use trial and error.

Give it a try, you need to cover the original set (polygon outline of the leaf) with copies of itself. You want to cover as much as possible with little overlap among the copies. Start by positioning and resizing the transformation that you already have (see figure 6). To move the transformation simply click anywhere in side the figure, hold and drag. To resize the figure click and drag on any of the nine small rectangles (handles) and drag to the desired size.

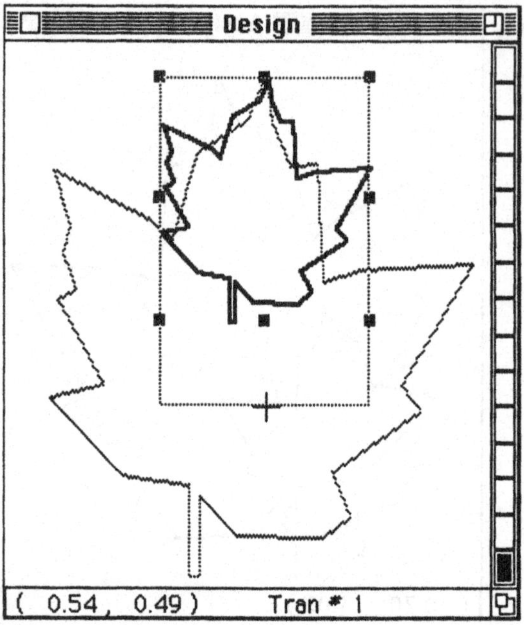

Figure 5.6 Moving and Resizing Transformation 1

Now add a second transformation by choosing **New Transformation** under the **Design** menu. Resize, rotate, shear and move the transformation to cover the right hand side. To rotate press the option key then drag one of the corner handles, to shear press the option key and drag a middle [side] handle. All of the commands are undoable (choose **Undo** under the **Edit** menu) so experiment.

Figure 5.7 Covering the right hand side.

Figure 5.8 Covering the left side.

Repeat the above steps to get a transformation to cover the left side of the leaf.

Finish the collage by adding two more transformations. Fill in the empty spot in the middle with a small transformation (figure 5.9). Make the stem by placing a very small copy in the bottom of the window (figure 5.10).

Figure 5.9

Figure 5.10

As you were adding and manipulating the transformations ***Fractal Attraction*** was calculating the corresponding equations. Bring the **code** window to the front to examine the equations that resulted from your efforts. Yours will undoubtedly be different from the code given here. If you have done a good job with the collage this set of numbers will reproduce the leaf. Do you see the significance? Thirty numbers are enough to generate a picture of the leaf. You could transmit these numbers to anyone who has a IFS program and they would be able to recreate your leaf.

1. $$\begin{bmatrix} x_{new} \\ y_{new} \end{bmatrix} = \begin{bmatrix} 0.500 & 0.000 \\ 0.000 & 0.656 \end{bmatrix} \begin{bmatrix} x_{old} \\ y_{old} \end{bmatrix} + \begin{bmatrix} 0.278 \\ 0.356 \end{bmatrix}$$

2. $$\begin{bmatrix} x_{new} \\ y_{new} \end{bmatrix} = \begin{bmatrix} 0.258 & 0.414 \\ -0.226 & 0.519 \end{bmatrix} \begin{bmatrix} x_{old} \\ y_{old} \end{bmatrix} + \begin{bmatrix} 0.471 \\ 0.225 \end{bmatrix}$$

3. $$\begin{bmatrix} x_{new} \\ y_{new} \end{bmatrix} = \begin{bmatrix} 0.295 & -0.351 \\ 0.230 & 0.626 \end{bmatrix} \begin{bmatrix} x_{old} \\ y_{old} \end{bmatrix} + \begin{bmatrix} 0.224 \\ 0.043 \end{bmatrix}$$

4. $$\begin{bmatrix} x_{new} \\ y_{new} \end{bmatrix} = \begin{bmatrix} 0.490 & 0.084 \\ -0.077 & 0.427 \end{bmatrix} \begin{bmatrix} x_{old} \\ y_{old} \end{bmatrix} + \begin{bmatrix} 0.191 \\ 0.092 \end{bmatrix}$$

5. $$\begin{bmatrix} x_{new} \\ y_{new} \end{bmatrix} = \begin{bmatrix} 0.040 & 0.000 \\ 0.000 & 0.242 \end{bmatrix} \begin{bmatrix} x_{old} \\ y_{old} \end{bmatrix} + \begin{bmatrix} 0.343 \\ -0.102 \end{bmatrix}$$

Figure 5.11

Step 5. See how well the IFS code approximates the leaf.

Make the fractal window active, choose **Run** and click on the **OK** button to see the result. Hopefully you did better than this attempt! If not, go back to the **Design** window and give it another try.

Figure 5.12

§6 Color Rendering Example (Color Mac Only)

This section will discuss using the *Fractal Attraction* options to color the generated fractal. It is assumed that the reader will be using a color Mac as they work through this tutorial. It is also assumed that **only the Random Algorithm** is being used. There is a different set of rules that apply for the deterministic algorithm.

Fractal Attraction can use color to signify two things. The first is to illustrate which transformation was used. Each transformation is numbered (see the code window). When a point (pixel) is plotted in the **Fractal** window the color is based on the transformation number that moved the previous point to this point. [You will find an interesting variation implemented in Fractal Attraction, the idea is to base the color on the transformation that landed on the previous point.]

The second way color can be used is based on the number of times a point has been "hit", referred to as color rendering. The user sets a first color, last color and the number of shades to use. *Fractal Attraction* computes a color for each point based on the number of times the point was hit. It does this by first finding the maximum number of hits any point had. It assigns the first color to points (pixels) that have been hit only once, the last color to points that have been hit the maximum number of times. For the points in between a linear interpolation is done and the point is assigned one of the shades between the first and last color. This will be explained in more detail in the second example.

Two examples will be used to illustrate the color features. In the first example color will be based on the transformation number. The second demonstrates color based on the frequency that points have been hit.

Example 1. (Color based on transformation number)

Start a new session with *Fractal Attraction.* Close the first window (untitled) you see and load the file called Sierpinski triangle from the IFS examples folder. Bring up the **Fractal** window and choose **Do 10000** from the **Fractal** menu. You should see a multicolored Sierpinski triangle in front of you. One part of it should be red, another green and the last part blue. Bring up the **Design** window, you should see three squares, a red, a green and a blue. In the **Fractal** window each point is colored the same color as the last transformation that hit that point. Click on the blue square in the

Design window, it now becomes highlighted and there will be a message at the bottom of the design window telling you that you have chosen transformation three. All of the points that are colored blue in the **Fractal** window, are blue because of transformation three (the blue transformation).

You can change the color of a transformation by selecting the transformation in the **Design** window then choosing **Select Color** under the Design menu. There is a nice shortcut to selecting a transformation and choosing a color: Along the right side of the **Design** window are three small rectangles that are the same color as the transformations. To select a transformation merely click in the appropriate small rectangle. For example, click in the green one, the green transformation is now selected. To bring up the dialog box that allows you to specify the transformation's color, double click in the corresponding small rectangle. Try it, double click in the red rectangle, the red transformation is now selected and you get a dialog box that allows you to specify a new color. Try changing to yellow, click on yellow in the color circle and then click on the **ok** button. Now go back to the **Fractal** window and choose **Do 10000** again, you may want to clear the picture first by selecting **Clear** under the **Edit** menu.

There is an interesting twist to this scheme that you are encouraged to experiment with. Make the **Fractal** window the front window and select **Transformation History** under the **Fractal** menu. Keep holding the mouse down and another, second menu will appear. In this submenu select **Back 1.** Clear the screen (clear under edit) and **DO 10000** again. Try the same thing with **Back 2** to see if you can figure out what is happening. Good Luck.

Example 2 (Color Rendering based on number of hits)

This example will color a fern based on the number of times each point is hit. Start a fresh session of *Fractal Attraction*. It is assumed that your computer is set to show at least 256 colors. Close the untitled (code) window and open the file **fern** in the IFS example folder. Bring up the **Fractal** Window and choose **Run** under the Fractal Menu. A dialog will appear asking for the number of transformations, type 60000 then click on the **OK** button.

Once the fern is generated you are ready for color rendering. *Fractal Attraction* has kept track of the number of times each point was hit. Take the naive approach and use the default values for the first try at color rendering.

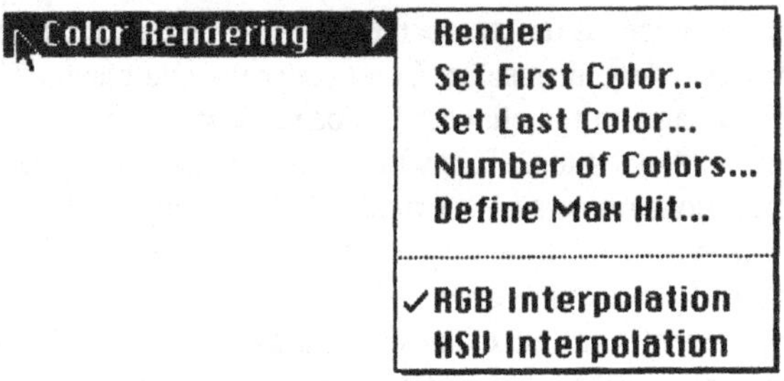

Drag down to **Color Rendering** (under the **Fractal** Menu) a submenus will appear, select **Render**. At the bottom of the window is a status bar. As *Fractal Attraction* calculates colors there will be a percentage providing a progress report. Eventually you should get a green fern that is red at the tips. The red points were hit the most often, the green the least often (but at least once!).

Perhaps you want a fern that goes from light green to dark green. This can be done by resetting the last color to dark green. Under the **Color Rendering** submenu is another choice called **Set Last Color**, choose this item. You will get a dialog called the color picker. In this dialog specify a color for the last color. You can specify the color either by clicking in the color wheel or setting the values directly. For this example enter the color as red, green and blue values directly. In the lower left of the color picker dialog set the red to 0, green to 24,000 and blue to 0. Click on the **OK** button. You should now have dark green as your last color. The current colors are shown on the right hand side of the **Fractal** window. Select **Render** again under the **Color Rendering** submenu (you do not need to run the random algorithm again to render again). You should now have a green fern that is dark green at the tips.

You might prefer a fern that has a more gradual change in the colors. To accomplish this set the number of colors to a higher value. The default value is three (which you can see by the rainbow in the right of the Fractal Window), change to ten colors. One way to change the number is by choosing **Number of Colors** under the **Color Rendering** submenu but this is a good chance to introduce another shortcut. Double click in the middle of the colors (rainbow) shown on the right side of the **Fractal** Window. This will bring up a dialog box to set the number of colors, type in ten. [By the way, double click in the bottom 1/3 of the rainbow to set the first color, top 1/3 to set the last color, and middle 1/3 to select the number of colors.] Click on the **OK** button and you should now see ten shade from light to dark green on the right side of the window. Select **Render** again and check the results.

The shading may not be as dramatic as you would wish. To improve it one can try doing more iterations. Before doing more iterations copy the current image to the clipboard for later comparisons. Make sure the **Fractal** window is the front window, then select **Copy** under the edit menu. There is now a copy of the image on the clipboard. For more iterations select **Run** again and click on the **OK** button. This will compute 60,000 more iterations. Once the program is done running try the render command again. To compare against the old image choose **show clipboard** under the edit menu and put the two windows side by side (clipboard and fractal).

You can experiment with all of these features, first color, last color, number of colors and number of iterations to get an interesting shading. Another parameter that affects the shading is the probability associated with each transformation which you can edit in the code window. This example will consider one more parameter, Max Hit.

When you choose render, *Fractal Attraction* scans all the points in the **Fractal** window to determine the largest number of hits. It uses this number in calculating the shading. The problem with this approach is that one point that has had numerous hits can skew the statistics. As a way around this Fractal Attraction will let you set a number called Max Hit . Any point that has more hits than Max Hit will be considered to have a value of Max Hit. You can set this number by selecting **Define Max Hit** under the **Color Rendering** sub-menu. Try it with the fern. First choose **copy** so that you can later compare. Now choose **Define Max Hit,** it defaults to 15, try changing it to 10. Select **Render** again and compare the results to the clipboard.

As you can see the color aspects of iterated function systems can be fascinating. Hopefully this section gave you enough information to go out and explore. If while you are exploring you invent a fascinating fractal the authors would love to see it. It may even be included, with proper credit given of course, in the library of fractal images the authors are compiling.

§7. The Fractal Attraction Application

This is the *Fractal Attraction* reference section, an overview of the windows is given and the menu items are explained in detail.

§7.1. The Windows

Fractal Attraction has three windows for working with an IFS code. The first is the window which contains the IFS code itself (the actual numbers which produce the image) called the **Code** window (you give it your own file name). In this window you can specify the number of transformations, the parameter values for each transformation and the coordinate system (minimum and maximum values for x and y). The **code** window displays the title of the file you are working with. It allows you to display and edit your code in three different formats: as matrix equations, as just the parameters of the matrix equations, or in scale/rotation format (see appendix b).

The second window is titled **Design**. In this window you can see the effect of each transformation on the gray square that forms the border (the initial set). It is titled **Design** because you can make changes in the IFS code from this window. Using the mouse you can add, remove, re-size, rotate, shear and move the transformations [See the section **Using the Design Window**].

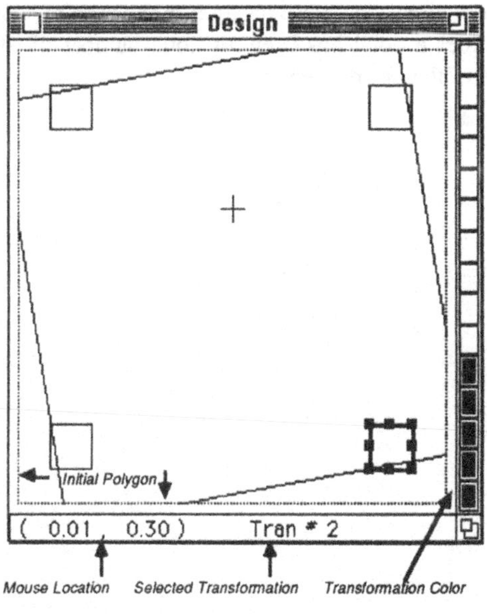

The third window is titled **Fractal**. In this window you will be able to apply the IFS code and see the results. You have a choice of two algorithms to generate your fractal, random or deterministic. The coordinate system can be changed by using the **Zoom In** and **Zoom Out** features. Once you have generated an image you can copy it, or save it to a file, for inclusion in other applications, such as **MacPaint.**

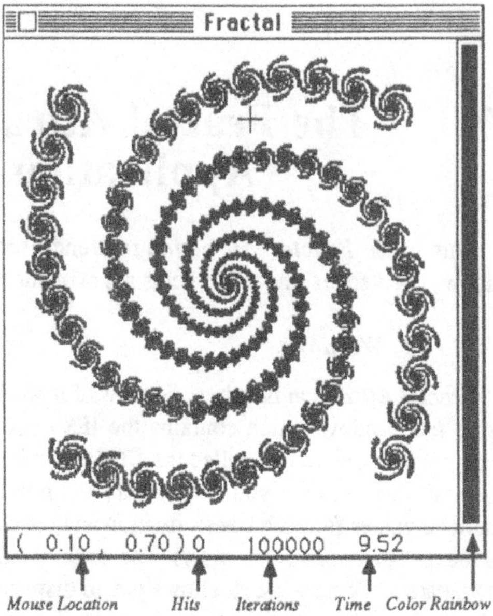

There are several numbers at the bottom of the window that provide status information: the mouse location in the current coordinate system; the number of times a point has been hit; total number of iterations; and the execution time of the last request.

§7.2. *The Menus*

The File Menu

New Creates a new, untitled code window (for entering an IFS code). Only allowed if there is not a code (IFS code) document already open

Open... Allows the user to open an existing IFS document. Only one document may be open at a time.

Save Saves the current IFS (code) document, including the initial polygon. The document is saved in a text format allowing it to be opened by any program capable of reading text. To examine the structure of a *Fractal Attraction* file open it using a word processor.

Save As Saves the current document with a different name. If the **Fractal** window is the active window choosing this item allows you to save the fractal image as a PICT file.

Close Closes the active window. If the active window is the edit window it will close the file. If the **fractal** or **design** window is the active window, the window will simply be hidden. To get the **Fractal** or **Design** window back choose them from the **Window** menu.

Page Setup... Brings up a dialog that allows you to specify page parameters for printing.

Print... Prints the active window.

The Edit Menu

Undo Allows the user to undo the effects of some commands.

Cut Standard Macintosh command for editing text.

Copy Copies text if the **code** window is active. If the **Fractal** window is the active window you can copy all or part of the picture. To copy just part of the picture use the mouse to select the portion to be copied. It is also possible to copy the **Design** window, just make it the active window.

Paste Paste text in the **code** window for standard Macintosh text editing.

Clear Works as standard text item in **code** window. When used in **Design** window it clears the selected transformation. If the **Fractal** window is the active window this command clears the screen.. Note that it is undoable for both the design and fractal windows (see **undo** under the edit menu).

Show Clipboard Shows the current content of the clipboard.

The Code Menu

Calc. Probabilities Calculates the probabilities (frequency) of applying each transformation based on the value of the determinant (*ad —bc*). If the determinant equals zero the probability defaults to 0.01.

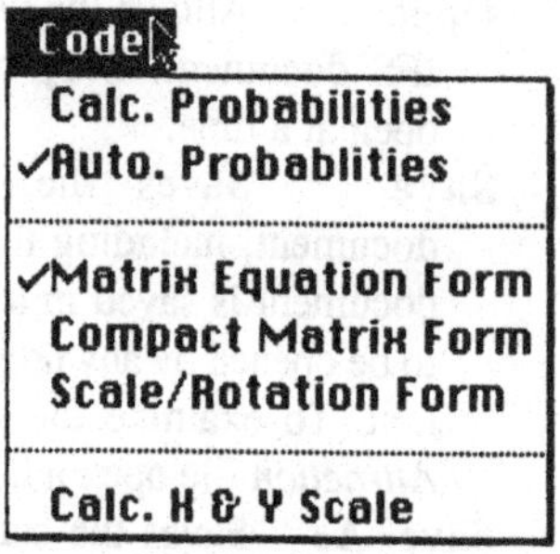

Auto. Probability If this item is checked the probabilities are automatically calculated when the random algorithm is used, if they do not already sum to 1.

Calc. X & Y Scale Determine the coordinates of the fractal (minimum and maximum *x* and *y*).

Matrix Equation Form If checked display the parameters in equation form..

Compact Matrix Form Show the transformation parameters in a compact, spreadsheet like form.

Scale/Rotation Form Show the parameters in the alternate form of scalings, rotations and translations (see appendix b for more information).

The **Design** Menu

These menu items are only available if the **design** window is the active window.

New Transformation
Add a new transformation to the **design** window and corresponding row to the **code** window. The initial values for the new transformation are: a = 0.5, c = 0.5, b = d = e = f = 0.

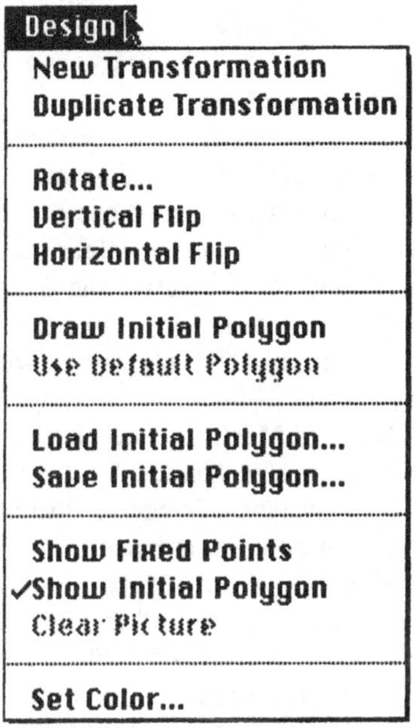

Duplicate Transformation
Duplicate the parameters for the selected transformation except offset slightly.

Rotate... Rotate the selected transformation by the given degrees. The number of degrees is specified in a dialog box. The specified degrees must be in the range -360 to 360.

Vertical Flip Reflect the y-coordinates across a horizontal line through the center of the bounding rectangle.

Horizontal Flip Reflect the x-coordinates across a vertical line through the center of the bounding rectangle.

Draw Initial Polygon Allows the user to draw a polygon in the **Design** window. The transformations are then applied to this new polygon (set of points). The polygon is limited to thirty points.

Use Default Polygon Replaces the user specified polygon by the default polygon, a rectangle that outlines the window.

Load Initial Polygon... Load the initial polygon from a disk file. The polygon is saved as a list of ordered pairs. To see the format of a file open with any text editor.

Save Initial Polygon... Save the initial polygon to a disk file. This allows you to save a copy of the initial polygon separate from the IFS code.

Show Fixed Points If checked, the unique fixed point for each transformation will be shown in the **Design** window.

Clear Picture Clear the picture (graphic image, e.g.maple leaf) that has been pasted into the design window.

Set Color Brings up a dialog box to set color of current transformation.

The Fractal Menu

The following items are available only if the **Fractal** window is active.

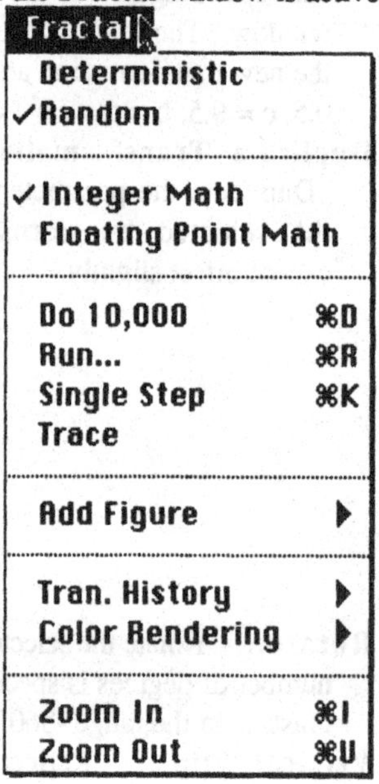

Deterministic This menu item along with the next one determines which of the two algorithms are used for computing the fractal. If this item is checked the deterministic algorithm will be used.

Integer Math If checked use integer arithemitic for the random algorithm. Integer arithemitic is faster but less accurate.

Floating Point Math Use floating point arithemitic with the random algorithm. Floating point is more accurate than integer, but slower. It is recommended that you only use floating point if your Mac is equipped with a floating point co-processor. Floating point math may be necessary for extreme zooms.

Random Use the random algorithm for plotting the fractal if checked.

Do 10000 or Do One This menu item will vary depending on whether the **Random** or **Deterministic** algorithm is selected. If the menu item is **Do 10000** , 10000 iterations of the random algorithm will be executed. If it is **Do One** it will do one set of transformations using the **Deterministic** algorithm

Run... Similar to **Do One** or **Do 10000** except that it lets you specify the exact number of iterations through a dialog box.

Single Step Does a single iteration of the random algorithm.

Trace Runs the random algorithm displaying the last point and providing information on the transformation used.

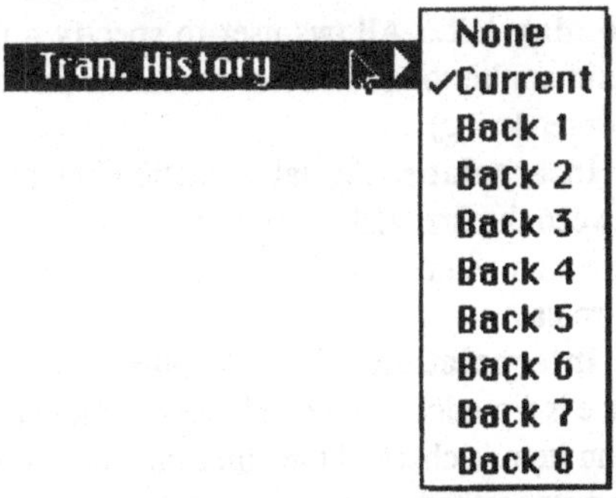

Tran. History (Color Mac only) Brings up a submenu that lets you decide what transformation to base the color on as the attractor (fractal) is being generated (random algorithm only). E.G. if **Current** is checked each new point will be plotted in the color of the transformation that landed on this point. If **Back1** is chosen the color will be based on the transformation one back, **Back2** is based on two transformations back, and so on. This is related to the idea of each point having an "address". The best way to understand this option is to see it in action. Bring up the Siepinski example file and try the various history options as you run the random algorithm.

If **None** is checked the attractor will be colored with just black, the advantage is that it is slightly faster.

Color Rendering (Color Mac only) Displays submenu for rending command and setting color rending options.

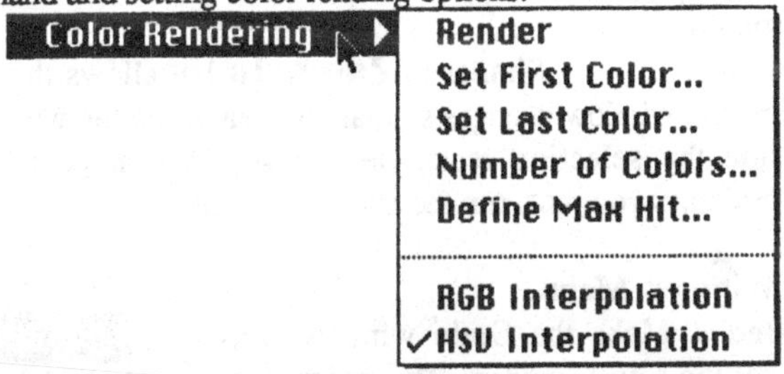

Render Calculate color of fractal points based on the number of times hit. See section on color rendering for more information.

Set First Color... Brings up color picker dialog which allows user to specify the first color for the fractal window.

Set Last Color... Brings up color picker dialog which allows user to specify the last color for the fractal window.

Number of Colors... Presents user with a dialog to enter number of colors to use in color rendering. Colors are interpolated values from first color to last.

Define Max Hit... Allows user to specify a number for the maximum hit to use in color rendering (see the section on color rendering).

RGB Interpolatin Fractal Attraction interpolates the colors between the first and the last. If this item is checked Fractal Attraction will use the Red/Green/Blue color space model to interpolate.

HSV Interpolation An alternative to RGB color space is a model based on Hue/Saturation/Value. If this menu item is checked the color interpolation will use this alternate method.

Add Figure ... Displays submenu for adding figures to the **Design** window. These commands are useful for adding an initial figure for using the deterministic algorithm.

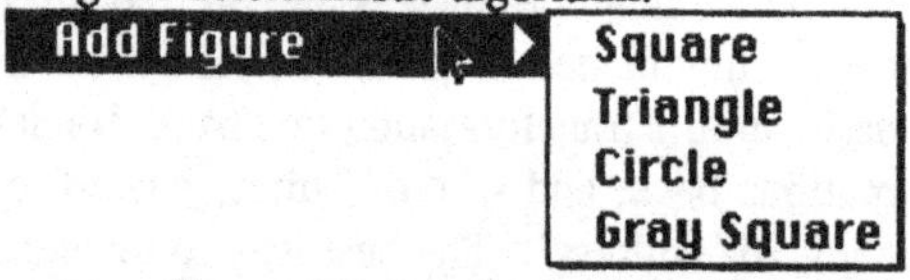

Square Draws a square in the **Fractal** window.

Circle Similar to **Square** but draws a circle.

Triangle Draws a triangle.

Gray Square Produces a filled (gray) square.

Zoom In Enabled if part of the **Fractal** window is selected. Zooms the window so the selection rectangle fills the window (changes the coordinate system). The image must be recalculated to see the higher resolution.

Zoom Out Similar to **Zoom In** but allows the user to step back; the window changes so that the area inside the window now fits inside the selection rectangle. To see the image with the new coordinates you must run the algorithm again.

The Window Menu

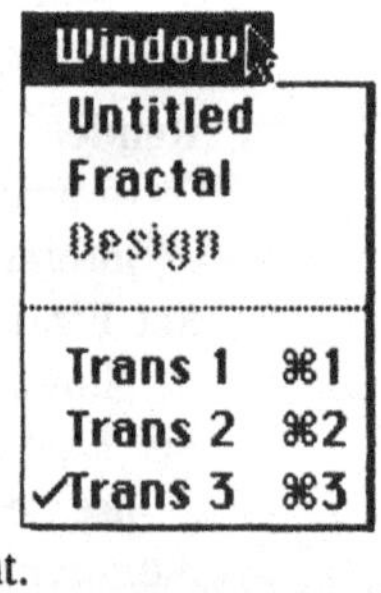

Untitled Make the **Code** window (IFS code) the active window. This menu item will change to match the title of the current code window.

Fractal Make the **Fractal** window active.

Design Bring the **Design** window to the front.

Trans 1 The behavior of the **Trans** commands vary depending on which window is active. If the code window is active it will be grayed out (It was put here because of the available space in this menu and since it can be used from either the **Design** or **Fractal** window).

If the **Design** window is the active window choosing this item will select transformation one. This is useful if transformation one is covered by other transformations making it difficult to select with the mouse.

If there is an open **code** window and **Fractal** is the active window the user may apply just a single transformation, in this case the first transformation. If the deterministic algorithm is chosen, the transformation is applied to the entire figure. For the random algorithm it will do one step using transformation 1.

Trans 2- Trans 15 These menu items will be added as you add transformations to the **code** window. The behavior is identical to **Trans 1** but uses the specified transformation.

§8. Literature Cited

Barnsley, M. *Fractals Everywhere,* Boston: Academic Press, 1988.

Devaney, Robert L. *An introduction to Chaotic Dynamical Systems,* 2nd ed. Menlo Park: Addison-Wesely, 1989.

Devaney, Robert L. *Chaos, Fractals and Dynamics. Computer Experiments in Mathematics,* Menlo Park: Addison-Wesely, 1990.

Edgar, Gerald A. *Measure, Topology, and Fractal Geometry,* New York, Springer-Verlag, 1990

Glieck, J. *Chaos: Making a New Science,* New York: Viking, 1987.

Mandelbrot, B. *The Fractal Geometry of Nature,* San Francisco: W. H. Freeman and Co., 1982.

Pitgen, H.-O, Richter,P.H., eds. *The Science of Fractal Images,* New York: Springer-Verlag, 1988

Pitgen, H.-O, Richter,P.H., *The Beauty of Fractal Images,* New York: Springer-Verlag, 1986

Appendix A: IFS Examples

The IFS codes for the following pictures reside in the IFS Examples folder.

Spiral_4

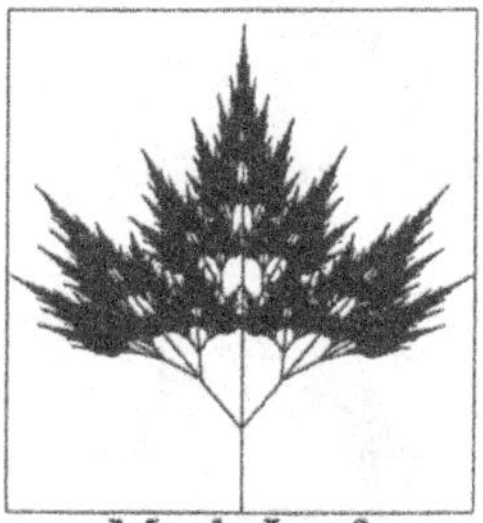

Maple Leaf

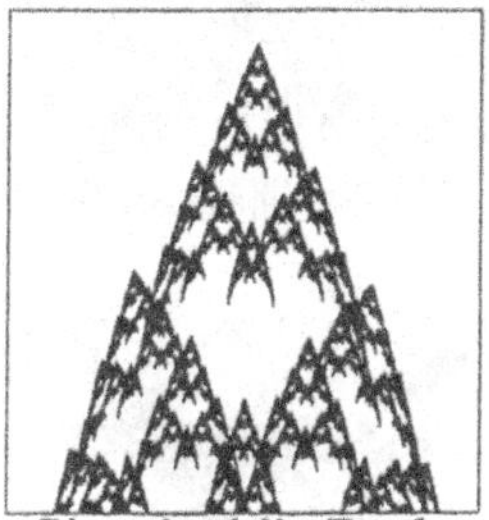

Sierpinski's Peak

Fern

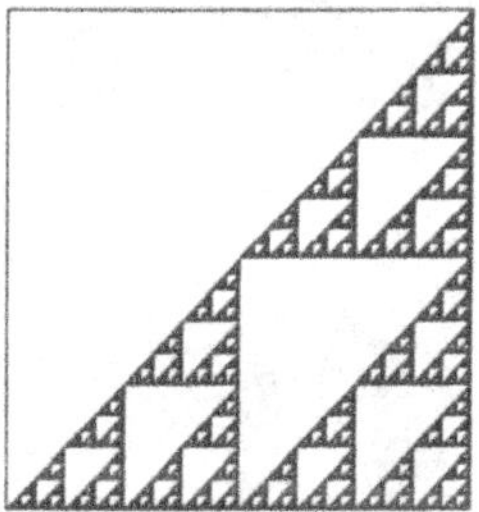

Sierpinski Triangle

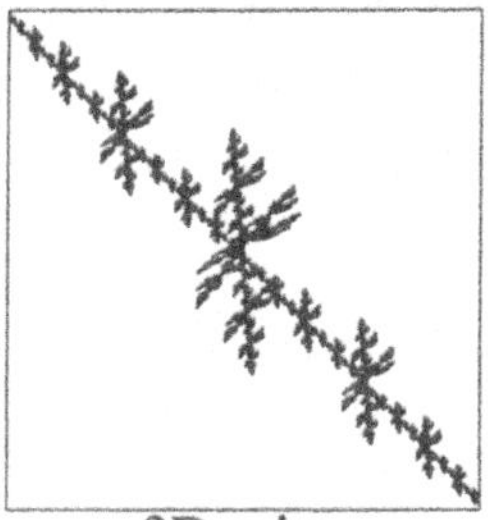

3D-wire

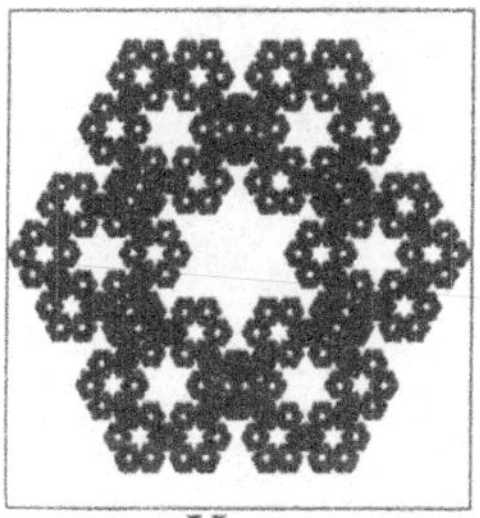

Hex

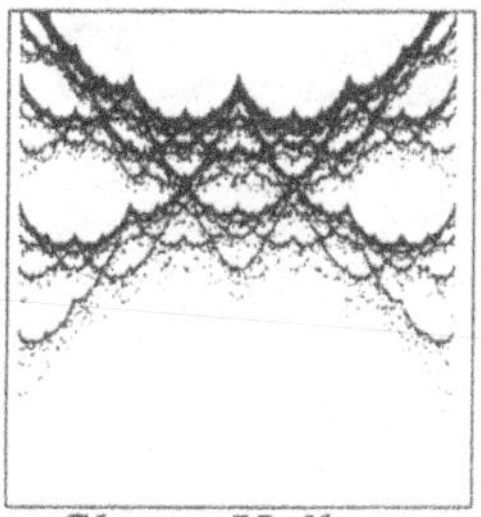

Sleepy Hollow

Kangaroos

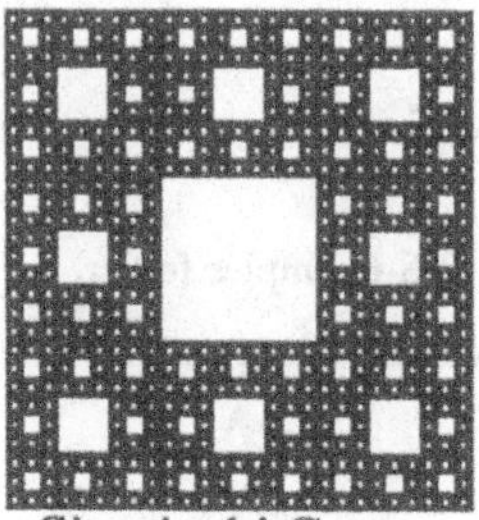

Sierpinski Carpet

Wheat

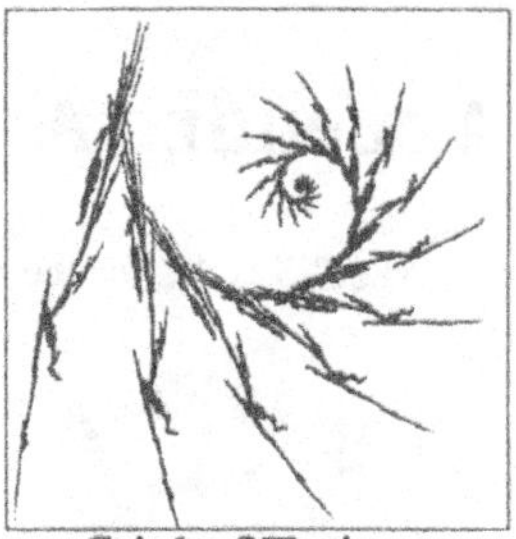

Stick_2Twist

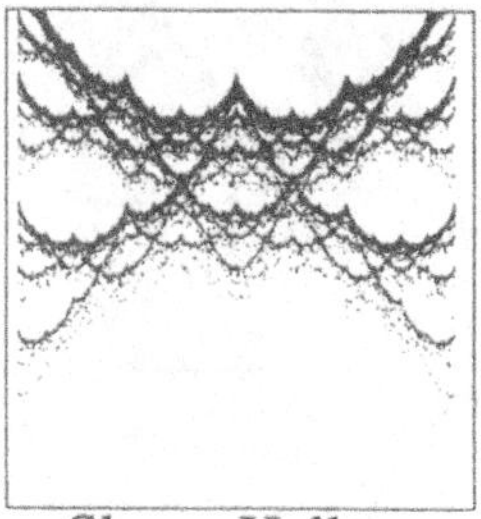

Sleepy Hollow

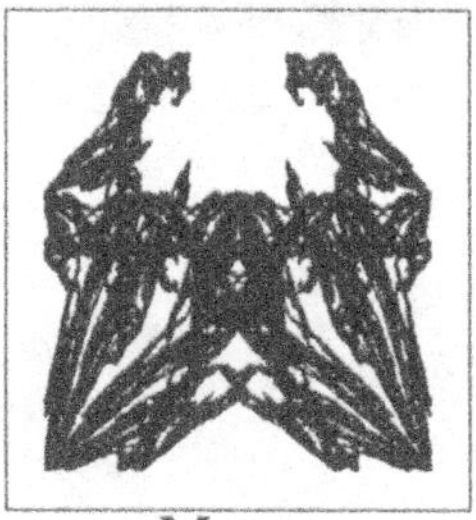

Mutant

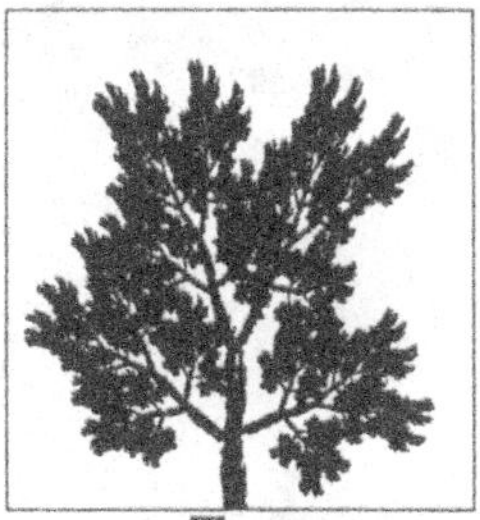

Tree

Eiffel Tower

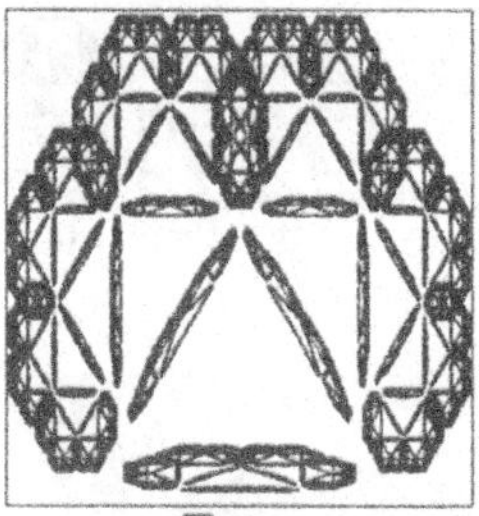

Face

Pine_Tree

Appendix B: Rotation, Scale, & Translation Form

Fractal Attraction normally works with affine transformations in the form of equation 1.

$$W\begin{pmatrix} x \\ y \end{pmatrix} = \begin{pmatrix} a & b \\ c & d \end{pmatrix}\begin{pmatrix} x \\ y \end{pmatrix} + \begin{pmatrix} e \\ f \end{pmatrix} \qquad (1)$$

Another form is to specify these transformations as rotations, scalings and translations. Equation 2 is equivalent to equation 1 where q and y are rotations, r and s are scalings and e and f are translations all of x and y respectively.

$$W\begin{pmatrix} x \\ y \end{pmatrix} = \begin{pmatrix} r\cos q & -s\sin y \\ r\sin q & s\cos y \end{pmatrix}\begin{pmatrix} x \\ y \end{pmatrix} + \begin{pmatrix} e \\ f \end{pmatrix} \qquad (2)$$

Fractal Attraction , in a concession to those who think scale, rotation, and translation is easier to visualize, will let you work in either form. To change forms you simply need to select the form you want under the **code** menu. You actually have three choices, the first two deal with *a,b,c,d,e,* and *f* directly. The third choice lets you work with scales and rotations.

Choosing **Matrix Equation Form**

1. $$\begin{bmatrix} x_{new} \\ y_{new} \end{bmatrix} = \begin{bmatrix} 0.177 & 0.354 \\ 0.177 & 0.354 \end{bmatrix}\begin{bmatrix} x_{old} \\ y_{old} \end{bmatrix} + \begin{bmatrix} 0.000 \\ 0.000 \end{bmatrix}$$

Choosing **Compact Matrix Form**

	a	b	c	d	e	f
1	0.177	0.354	0.177	0.354	0.000	0.000

Choosing **Scale/Rotation Form**

	Scale		Rotation		Translation	
	x	y	x	y	x	y
1	0.250	0.500	45.0	-45.0	0.000	0.000

In equation (2) r is often called the x-scale, s the y-scale, q the x-rotation, y the y-rotation, and e and f are called the x and y translations, respectively. This is the same representation *Fractal Attraction* uses for **Scale/Rotation Form**.

Fractal Attraction internally stores and saves the numbers as the parameters *a,b,c,d,e,* and *f*. When displaying scale/rotation form it simply calculates the corresponding scales and rotations. When you change an angle or scale *Fractal Attraction* calculates the corresponding matrix parameters

(a,b,c,d). This leads to one idiosyncrasy (bug?) when working with Scale/Rotation Form. If you try to set the rotation and the corresponding scale (for example trying to set the x rotation and the x scale is 0) the rotation will always end up being zero. In other words, you need to have a non-zero scale to set the corresponding rotation.

Appendix C: Changing Fractal Window Size

The size of the Fractal Window in the current version is fixed to one size at startup, usually 256x256 pixels (~ 3.5" x 3.5"). In this appendix we will explain how to change the size of the startup fractal windows to one of two possibilities, 384 x 384 or 512 x 512. The drawback to increasing the fractal window size is the amount of memory needed and the need for a bigger monitor. The need for memory also increases with the screen depth (i.e. 8 bit or 24 bit color).

The following table indicates recommended memory sizes for the various options.

	1 bit	8 bit	24 bit
256 x 256	600Kb	750Kb	1024 Kb
384 x 384	750Kb	1450Kb	2300Kb
512 x 512	1024 Kb	2048Kb	3400Kb

To change the size of the fractal window you must be running Fractal Attraction. Under the edit menu you will find an item titled **Set Startup Size.** Choosing this item will bring up a dialog box. Simply click on the size of the fractal window you desire. The changes will not take effect until you quit the program and start it up again.

If you are running under multifinder you may need to change the size of the memory partition allocated to Fractal Attraction. This is done while in the **Finder**. Select the Fractal Attraction Icon (click once), then from the **File** menu select **Get Info**. In the resulting dialog box change the **Application Memory Size** to the appropriate number from the above table.

Made in the USA
Monee, IL
13 September 2026

62977015R00046